Global Issues in Water Policy

Volume 23

Editor-in-chief
Ariel Dinar, Department of Environmental Sciences, University of California, Riverside, California, USA

Series editors
José Albiac, Department of Agricultural Economics, Unidad Economia, CITA-DGA, Zaragoza, Spain
Stefano Farolfi, CIRAD UMR G-EAU, Montpellier, France
Rathinasamy Maria Saleth, Chennai, India
Guillermo Donoso, Department of Agricultural Economics, Pontificia Universidad Católica de Chile, Macul, Chile

More information about this series at http://www.springer.com/series/8877

Slim Zekri
Editor

Water Policies in MENA Countries

Springer

Editor
Slim Zekri
CAMS, Department of Natural Resource Economics
Sultan Qaboos University
Al-Khod, Sultanate of Oman

ISSN 2211-0631   ISSN 2211-0658   (electronic)
Global Issues in Water Policy
ISBN 978-3-030-29273-7   ISBN 978-3-030-29274-4   (eBook)
https://doi.org/10.1007/978-3-030-29274-4

© Springer Nature Switzerland AG 2020
This work is subject to copyright. All rights are reserved by the Publisher, whether the whole or part of the material is concerned, specifically the rights of translation, reprinting, reuse of illustrations, recitation, broadcasting, reproduction on microfilms or in any other physical way, and transmission or information storage and retrieval, electronic adaptation, computer software, or by similar or dissimilar methodology now known or hereafter developed.
The use of general descriptive names, registered names, trademarks, service marks, etc. in this publication does not imply, even in the absence of a specific statement, that such names are exempt from the relevant protective laws and regulations and therefore free for general use.
The publisher, the authors, and the editors are safe to assume that the advice and information in this book are believed to be true and accurate at the date of publication. Neither the publisher nor the authors or the editors give a warranty, expressed or implied, with respect to the material contained herein or for any errors or omissions that may have been made. The publisher remains neutral with regard to jurisdictional claims in published maps and institutional affiliations.

This Springer imprint is published by the registered company Springer Nature Switzerland AG.
The registered company address is: Gewerbestrasse 11, 6330 Cham, Switzerland

Contents

1 An Overview of the Water Sector in MENA Region 1
Slim Zekri and Aaisha Al-Maamari

2 Water Policy in Algeria 19
Nadjib Drouiche, Rafika Khacheba, and Richa Soni

3 Existing and Recommended Water Policies in Egypt 47
Khaled M. AbuZeid

4 Iran's Water Policy 63
Farhad Yazdandoost

5 Water Policy in Jordan 85
Tala Qtaishat

6 Oman Water Policy 113
Slim Zekri

**7 Water Resources in the Kingdom of Saudi Arabia:
Challenges and Strategies for Improvement** 135
Mirza Barjees Baig, Yahya Alotibi, Gary S. Straquadine
and Abed Alataway

8 Water Policy in Tunisia 161
Mohamed Salah Bachta and Jamel Ben Nasr

9 The Water Sector in MENA Region: The Way Forward 185
Slim Zekri

Index 201

v

About the Editor

Dr. Slim Zekri is Professor and Head of the Department of Natural Resource Economics at Sultan Qaboos University (SQU) in Oman. He earned his PhD in Agricultural Economics and Quantitative Methods from the University of Cordoba, Spain. He is Associate Editor of the journal *Water Economics and Policy*. He has worked as a Consultant for a range of national and international agencies on natural resource economics, policy and governance, agriculture, and water economics in the Middle East and North Africa. He is Member of the Scientific Advisory Group of the FAO's Globally Interesting Agricultural Heritage Systems. His main research interests are water economics and environmental economics. In 2017, he was awarded the Research and Innovation Award in Water Science from the Sultan Qaboos Center for Culture and Science.

Chapter 1
An Overview of the Water Sector in MENA Region

Slim Zekri and Aaisha Al-Maamari

Abstract The Middle East and North Africa region is experiencing a widening gap between freshwater supply and demand caused by population and economic growth and climate change. This book addresses water scarcity issues in the MENA region and gives an overview of the current water policies in seven MENA countries: Algeria, Egypt Iran, Jordan, Oman, Saudi Arabia, and Tunisia. This book includes an introductory chapter and seven chapters showcasing water policies in each country. This introductory chapter gives a quantitative representation and description of current available water resources; water demand for industrial, domestic, and agricultural purposes; and water per capita decline over time. The seven chapters provide details on the main challenges faced in each of the countries in the water sector. The chapters address the laws governing water use in the three economic sectors, water supply, water pricing and cost recovery and irrigation efficiency, and technology adoption. The increase of supply from non-conventional resources such as desalination and reuse of treated wastewater is analyzed. The chapters end up discussing how the countries are adapting to climate change and the role of research and innovations.

Keywords Water supply · Water demand · Water policy · Climate change · Research

1.1 Purpose of the Book

This book enters a virgin soil and contributes to the water sector policies in the Middle East and North Africa (MENA) region. A seminal work has been done by the World Bank (2017b), which considers MENA as a whole. This book is an important contribution about countries from a region of crucial importance for water management. Physical, natural, socio-economic, and political constraints make this region a sort of "laboratory" for water management around the world. This book has

---

S. Zekri (✉) · A. Al-Maamari
CAMS, Department of Natural Resource Economics, Sultan Qaboos University, Al-Khod, Sultanate of Oman
e-mail: slim@squ.edu.om

© Springer Nature Switzerland AG 2020
S. Zekri (ed.), *Water Policies in MENA Countries*, Global Issues in Water Policy 23, https://doi.org/10.1007/978-3-030-29274-4_1

contributions from seven countries. Each chapter discusses the same dimensions of the water sector in a given country, allowing the readers to compare among the countries. Although MENA countries have a lot of water problems in common, the way the challenges are addressed differs from one country to another. This book starts with an introductory chapter that gives an overview of the water situation in MENA. Seven chapters follow with detailed cases on specific challenges in each country as well as the policies implemented so far. A concluding chapter finally compares the different reforms undertaken in the seven countries, the successful experiences, and the challenges ahead.

1.2 Introduction

The Middle East and North Africa region (MENA) is formed by 21 countries. A number of these countries are currently in a state of political instability, conflicts, or wars, which resulted in human suffering, mass migration, severe damages to water infrastructure, and halt of public investments. The present book presents an overview of the water policy for the following seven countries: Algeria, Egypt, Iran, Jordan, Oman, Saudi Arabia, and Tunisia. The selected countries are representatives of Maghreb, Mashreq, and Gulf countries. The World Bank (2017b) report is a seminal work on MENA water policy. The report has drawn an accurate image of water scarcity in the region and spelled the required reforms both in the water sector and beyond it to avoid the failures and economic losses that would result from an unmanaged scarcity. Two recent World Bank (2017a, 2018a, b) reports presented an update of the state of water scarcity as well as the status of water security in the MENA region. They portrayed the existing challenges and chances to overcome the water security issues. The reports questioned the sustainability and efficiency of water resources management, the reliability and affordability of water services, and whether the water-related risks are appropriately recognized and mitigated. This introductory chapter gives a quantitative representation and description of current available water resources; water demand for industrial, domestic, and agricultural purposes; and water per capita decline over time.

1.3 Water Resources in MENA

Among all regions in the world, the MENA region, comprising 21 countries,[1] experiences high level of water scarcity and variability in available freshwater resources over time. Figure 1.1 shows the available renewable water per capita in MENA

---

[1] MENA countries: Algeria, Bahrain, Dijbouti, the Arab Republic of Egypt, the Islamic Republic of Iran, Iraq, Israel, Jordan, Kuwait, Lebanon, Libya, Morocco, Oman, Palestine, Qatar, Saudi Arabia, the Syrian Arab Republic, Tunisia, the United Arab Emirates, and the Republic of Yemen.

1 An Overview of the Water Sector in MENA Region

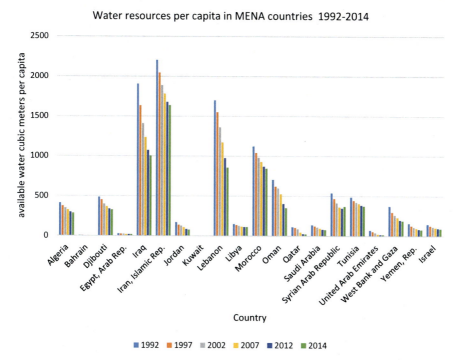

Fig. 1.1 Available renewable water per capita in MENA countries 1992–2014. (Source: World Bank 2018b. database, data for 2014)

countries for the years 1992–2014. The figure clearly shows that the available water per capita is dwindling over time for all MENA countries, reflecting an increased scarcity in the region. MENA countries are also characterized by high evapotranspiration rates, given the high temperatures most of the year. The common water sources for the MENA region are surface and groundwater in addition to non-conventional water such as desalination and recycled treated wastewater. Figure 1.2 shows that Iran, Iraq, Lebanon, and Morocco have relatively the highest quantities of renewable water resources compared to other MENA countries, with water level of more than 800 cubic meter per capita per year. On the other extreme, Gulf Cooperation Countries (GCC), Jordan, Libya, and Yemen have the lowest level of available renewable water resources per capita and they highly depend on depleting groundwater and desalination. Figure 1.3 shows the distribution of surface water and groundwater resources. GCC countries, Djibouti, Libya, and Jordan have the lowest quantities of groundwater and surface freshwater. Egypt, Iraq, and Syria depend on transboundary water resources such as rivers originating from other regions and aquifers which are shared with other countries. Rivers are the main water source used for crop irrigation in Egypt and Iraq, which are threatened by the increasing level of water use by upstream countries.

Population growth, fast economic growth, urbanization, and climate change are the main threats to the renewable freshwater resource (World Bank 2017a, b) besides

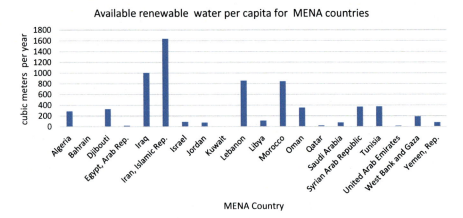

Fig. 1.2 Renewable water in m³ per capita per year. (Source: World Bank 2018b. database, data for 2014)

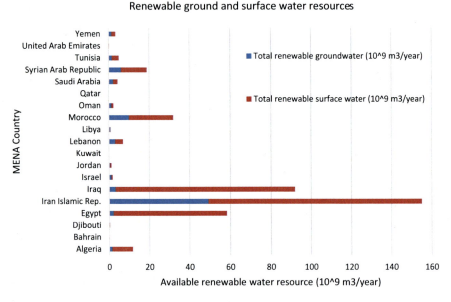

Fig. 1.3 Renewable volumes of groundwater and surface water in MENA. (Source: FAO AQUASTAT 2018, data for 2014)

to the international instability driving a number of MENA countries to increase food local production despite being non-competitive. The estimated population of MENA increased from 0.3 billion in 2000 to 0.4 billion in 2017 with a growth rate of 1.72%. It is expected that population will reach nearly 0.5 billion in 2030, considering a 1.2% average growth rate (World Bank 2018b). Demand for water resources will

1 An Overview of the Water Sector in MENA Region

increase for the agricultural, domestic, and commercial sectors, which will put higher pressure on the quantity and alter the quality of water resources in the region. Hence the imperative by MENA countries to reform their water policies and legislations and to efficiently allocate the water supply to fulfill current and prospect water demand. As shown in Fig. 1.3, most of the renewable water is in the form of surface water with a share of 78%. However, for most of the GCC countries, such as Bahrain, Kuwait, Qatar, and the United Arab Emirates (UAE), conventional water is exclusively in the form of groundwater with very low renewable volumes.

Figure 1.4 shows the current use of the available water resources based on sustainable and unsustainable types. All MENA countries depend partially on fossil water, unsustainable non-renewable groundwater. The most dependent is Libya with 80% of the water withdrawn coming from non-renewable groundwater, followed by Saudi Arabia with 75% of water used from non-renewable aquifers. The least exposed is Morocco with almost 2% of water withdrawals from unsustainable groundwater.

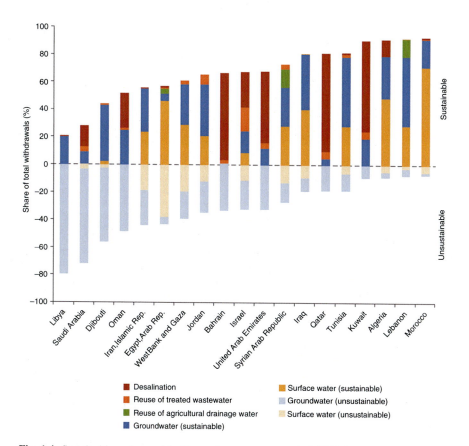

Fig. 1.4 Sustainable and unsustainable used water resources in MENA in percentage. (Source: World Bank 2018a, b)

1.4 Water Allocation

With continued noticeable increase in population and climate change, the percentage of water allocated to agricultural uses has increased in recent years. Figure 1.5 shows the percentage of water consumed by the agricultural, industrial and domestic sectors, respectively, for selected MENA countries. The figure shows that the largest quantities of renewable freshwater are used for agricultural purposes in almost all MENA countries, with an average use of over 70%. The domestic sector is second in terms of total freshwater demand. Finally, the industrial sector uses the lowest portion of freshwater resources with less than 6% of total uses.

Currently, groundwater is heavily used in all MENA countries because of the open access type of the resource and the availability of electricity for pumping in rural areas. Kuper et al. (2017) affirm that groundwater in North Africa will not provide a buffer capacity against drought and rainfall variability in the long term and cannot mitigate the effect of climate change as many suggest. This is because hard allocation choices are being postponed and short-term economic development and social welfare are favored at the expense of social and environmental sustainability of groundwater use. The authors assert that groundwater is being managed as a political "safety valve" where ambitious public policies encourage investments in high value crops, while perfectly knowing the resulting decline of the groundwater table. The increased variability of rainfall events and less spread of rain days is another reason for the higher abstraction rates of groundwater. In all MENA coun-

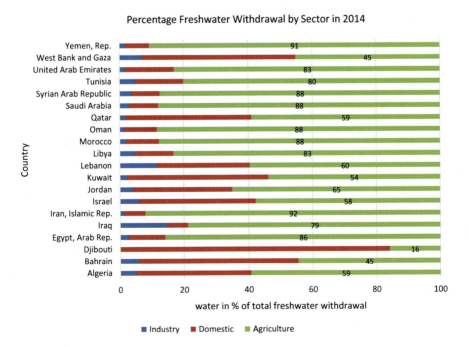

Fig. 1.5 Water allocation by sector in MENA. (Source: World Bank 2018b)

tries, except Jordan, there is no proper allocation of groundwater among users and in most cases there is no control of volumes abstracted. If groundwater policies are not properly addressed, drastic consequences on availability of groundwater for urban and domestic uses are expected, in the medium run, in several MENA countries. Groundwater over-abstraction can be reversed only if less water-intensive economic activities are considered conjointly with a proper mechanism of property rights definition and allocation. In other terms, industrialization, introduction of knowledge economy in rural areas, and creation of better paid non-agricultural jobs to maintain social peace seem unavoidable to trigger sustainable groundwater management policies. Currently, there is too much dependency on groundwater to alleviate poverty and maintain rural populations in place because of absence of alternative job opportunities. But current rural populations are leaving at water-credit, and rural exodus from several non-managed groundwater-depending areas should be expected in a number of MENA countries which will worsen the acute problems of large cities' suburbs.

Sanchez and Rylance (2018) reported more than 30 unrests and protests, in North African countries, related to water-supply-induced scarcity in 2015. The authors correlated the water service unrests to population size and growth, GDP, and GDP per capita for five North African countries. They showed that the frequency of water-related social unrests is higher during lower-than-average rainfall years, where population is dependent on limited water services. Lower-than-average rainfall years result in disruption of natural surface flows in the absence of institutional distribution.

Despite the fact that the agricultural sector is using the largest volume of water, all MENA countries, without an exception, have a food trade balance deficit (Fig. 1.6) that reached a total of \$ 29.6 billion in 2016 (WITS 2016). The trade balance is partly explained by the scarcity of water resources, but also depends on population size, income per capita, and food preferences. The food trade balance will probably worsen in the future given the high population growth and the expected decrease in available water resources. Furthermore, some of the current agricultural exports from MENA countries are based on non-renewable resources, like dates, for instance, from Algeria and Tunisia, which will reflect negatively on the export values in the future. Most of the MENA countries still have limited access to rich countries' markets where they can market high value products that require less water. Non-tariff barriers are limiting exports due to the presence of pesticide residues. On the positive side, the adoption of technologies that improve irrigation water efficiency and a responsible use of pesticides and chemicals would likely contribute to increase exports and better balance the food trade. These changes will occur only if market opportunities for agricultural products and logistics for export are enhanced. This will probably benefit the highly professional medium and large farmers who can master the production process, have access to funding, and are connected to networks but not the small farmers located in remote areas. From another perspective, however, as well documented by Chris et al. (2017), efficient irrigation technology will increase farmers' profit but will also result in higher agricultural water demand and competition for water among sectors. Hence the conflict

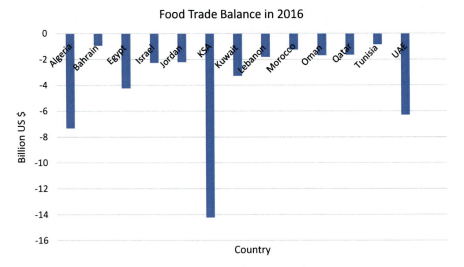

Fig. 1.6 Food trade balance for selected MENA countries in 2016. (Source: data from WITS 2016)

between reducing the food trade balance and saving agricultural water and allocating part of it to urban/environment uses. This poses the question if encouraging more food export from water-scarce countries is desirable. In all cases, in the absence of rural development strategies based on low-water-demanding industries, governments in developing countries are short of options. Overall, the alignment of agricultural production and food trade policy with water security goals seems in the right direction. However, recent food prices shocks, disruption of transport, conflicts, and wars are reviving old reflexes for food security that might have negative impacts on water security in the long run for certain MENA countries. Elmi (2017) justifies the concerns of food-import-dependent countries by the fact that international food prices are inversely correlated to oil prices and that the political and economic situations prevailing in the exporting countries affect the total supply of food available in the international market. Furthermore, the author stresses that 75–80% of the food entering the GCC countries passes through either Bab Al Mandab or Hormuz Straight, two passages affected by political instability and war. The author recommends the use of high efficient water technology such as hydroponics to improve the food security by increasing production in a sustainable fashion and improve the skills of the labor. He also recommends introducing food demand policies to curb down the food waste.

1.5 Recycling of Treated Wastewater

One of the water supply management targets by MENA countries is the potential use of large portions of treated wastewater for groundwater recharge, agricultural and landscape irrigation, the industrial sector, and to reserve the scarce renewable freshwater for domestic uses (WB 2017).

Figure 1.7 shows the average quantities of wastewater produced, collected, and treated by country for the period 2008–2012. The volume of wastewater collected in MENA countries represents 64% of the wastewater produced by the municipal sector. Similarly, the volume of wastewater treated represents only 43% of the total volume collected. In other terms, only 27% of the total volume of wastewater is being treated on average. This does have pollution implications on receiving water bodies. Furthermore, even though some countries have reached 80% of treatment of the collected water, they still have some pollution impacts given the fact that they treat only up to secondary level with treatment plants capacity being exceeded in several cases. The result is the pollution of beaches in big cities obliging citizens to travel long distances to find swimmable beaches. Algeria has been successful in reversing this pollution by introducing tertiary treated wastewater (TTWW) treatment and improving the quality of seawater for swimming in the capital's beaches (Drouiche and Soni chapter). The majority of Tunis capital beaches are not suitable

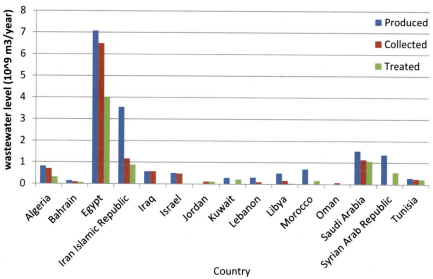

Fig. 1.7 Wastewater production, collection, and treatment in MENA. (Source: FAO AQUASTAT 2018, data for 2008–2012)

for swimming due to dumping of secondary treated wastewater in the Gulf of Tunis despite Tunisia being considered a success story in terms of wastewater management with 86% of households connected to the network (ONAS 2017). Large volumes of treated wastewater are also discharged into the Arabian Sea, a semi-enclosed gulf, resulting in accumulated pollutants threatening the marine environment (Van Lavieren et al. 2011) as well as the desalination process with harmful algae blooms. Further treatment of wastewater would come at higher costs that could be afforded by oil-rich countries, while non-oil MENA countries will unlikely invest in advanced treatments due to the absence of willingness of governments to increase the cost of service. The absence of studies on citizens' willingness to pay to protect beaches and other water bodies used for recreation does not help decision-makers see the potential benefits of advanced treatments.

Treated wastewater has also the potential to improve crops production if utilized efficiently. Figure 1.8 shows the percentages of volumes reused for irrigation purposes. On average, 37% of the treated wastewater is being reused for irrigation in MENA with high differences among countries. Some countries such as Jordan and Libya use 90% and 100%, respectively, of the treated wastewater. But overall most of the treated wastewater (63%) is discharged actually on surface water bodies or in the sea. The reasons for not reusing are multiple. WB (2017a) argues that a major obstacle to treated wastewater recycling is social acceptability. A close look at the experience of several MENA countries shows that barriers to wastewater recycling are linked to the price of treated wastewater, the degree of treatment and regulations, the rainfall uncertainty, and the farm size and crop mix. Wastewater demand is the highest where water scarcity is high, price of treated wastewater is competitive compared to conventional resources, and the quality of the treated wastewater is safe for

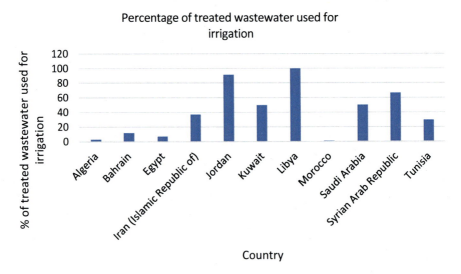

Fig. 1.8 Percentage of treated wastewater reused for irrigation in selected MENA countries. (Source: FAO AQUASTAT 2018, data for 2008–2012)

a large variety of crops. In Saudi Arabia, Al Hassa, tertiary treated wastewater is highly demanded given the absolute scarcity due to the groundwater level drawdown, salinity, and the existence of date plantations that require water. In fact, all the TTWW of the city is delivered to farmers free of charge. Even farmers who used to pump from the aquifers are changing to TTWW for irrigation given the high salinity of the groundwater (Zekri 2010). Another variable is the climate and type of crop. For instance, in northern Tunisia, farmers are very reluctant to use treated wastewater despite the low price. According to the regulations, the secondary treated wastewater can only be used to irrigate forage crops or cereals. The average rainfall in northern Tunisia is around 400 mm a year allowing rain-fed farming. Zekri et al. (2016) have shown that in Oman farmers are willing to use tertiary treated wastewater but the price proposed to farmers is double than their willingness to pay, resulting in a very low rate of reuse in the agricultural sector. In general, treated wastewater reuse for irrigation in MENA countries is either inefficiently used or the pricing system is inefficient. Treatment and distribution cost would not be fully recovered unless the applied pricing system and reuse rate are high enough to cover the maintenance and operation costs.

Given the population's spatial distribution, often the largest treatment plants are located in coastal areas while farms are located in the internal regions far from the treatment plants requiring huge investments in infrastructure and above all high operating and maintenance costs for pumping. In addition, most rural and internal areas of MENA countries have no sewage system or nearly of poor-quality sewage and sanitation infrastructure. So there is a need to improve the quality of the treated wastewater and build pipes network and sanitation system so that treated water can be reused for irrigation. Optimal location of treatment plants that takes into account where the treated wastewater is going to be recycled should be undertaken during the planning process. The possibilities of substituting good freshwater quality by treated wastewater will remain a dream given the fragmented water regulations and the lack of integrated solutions.

1.6 Urban Water

Most MENA countries have reliable piped urban water networks in the cities. However, according to LAS (2018), intermittence of supply of urban water is still an issue in many Arab countries such as Jordan where water is delivered once a week and Lebanon, Sudan, and Yemen with deliveries of 3–4 days/week. Residents usually adjust by storing water in rooftop tanks. The intermittence of supply is either due to water scarcity or due to the high costs of operation and maintenance of the system. In fact, maintaining a high pressure on the pipes 24 h a day is energy intensive and costly. Despite the fact that most connected people to piped water pay for water services based on a volumetric price, it is apparent that the water prices do not cover the costs and/or the non-revenue water represents an important portion of the water supplied. The perception that increasing water prices would cause social instability is one major obstacle to price reforms and cost recovery. In many MENA countries, urban water prices are terribly low that it is more expensive to buy one bottle of water from a supermarket than to pay for 1 m^3 from the network. This applies not only to Saudi Arabia and Kuwait among others but even for Tunisia, which has implemented a program of water cost recovery, but still has a very heavily subsidized first and second block tariffs. MENA countries still deal with the access affordability to water and equity problems with old pricing methods rather than exploring new methods that take into consideration the number of family members, indoor/outdoor uses, season, cost of service, environmental impacts, and the marginal cost of water. As a result, an important part of the subsidy ends up in the pockets of wealthy families. Indeed, despite the fact that water tariff might be tiered and progressive, the richest households are smaller in size compared to the poorest. Their consumption might fall in the first/second block tariff. For instance, in Tunisia, "…the richest quintile captured a greater share of the total subsidy (31%) as opposed to the poorest quintile which captured only 11%" (World Bank 2017a). Kotagama et al. (2016) estimated that wealthy families in Oman receive a water subsidy of $1846 per household/year. This shows clearly that the tiered pricing method is totally unfair, while populist voices and many decision-makers still sell it as a policy for protection of low-income classes. As well coined by Whittington (2016), overall subsidies are large, very poorly targeted to poor households, and water sold in the upper blocks is not high enough to generate revenues to cross-subsidize low-volume/low-income water users.

Water prices below cost result in high demand essentially if combined with high income. Figure 1.9 shows that water demand per capita is the highest in the water poor GCC countries. The consumption is above 400 l/cap/day where urban water

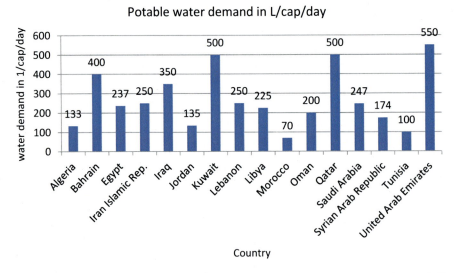

Fig. 1.9 Potable water demand in l/cap/day in the capital cities of MENA countries. (Source: Self collected from different sources and compiled)

supply depends mostly on expensive desalination process. In MENA only Morocco and Tunisia are in the range of 100 l/cap/day. The figure shows clearly the absence of water demand management in most MENA countries.

One more dangerous result of water prices below cost is that water utilities skimp on maintenance and differ network expansion (Abou Rayan and Djebedjian 2016). As far as water prices do not reflect scarcity, the supply and demand cannot balance. Policies cannot be just built on perceptions. Most MENA countries still lack reliable water economic and policy studies that would help decision-makers undertake necessary pricing reforms and overcome unjustified fears of political instability.

In terms of quality, access to potable water has been improving in MENA, especially in urban areas of the richer countries. Statistics show that access to improved drinking water in rural and urban areas of MENA countries increased slightly from 75.9% to 78.5% between 2003 and 2012. However, several rural areas still lack a reliable/quality supply of water, especially in the poorer countries with 70% of the rural populations concerned with the problem (Bryden 2017). Even in cities with intermittent supply that rely on rooftop storage, quality can be an issue when the tanks are made from asbestos. Recent studies show that asbestos fibers in water could pose unpredictable risks to humans after detecting asbestos in mice liver and blood, after 60 days of exposure (Zheng et al. 2018). Emerging pollutants is another concern that MENA countries should start looking at in order to improve the quality of delivered water (Teodosiu et al. 2018). This would require further investments in treatment technologies by water utilities which would also result in sensorial improvements requested by consumers. Consumers are turning to bottled water for taste as well as safety reasons, which indicates willingness to pay for better water quality from the taps. Thus, price reforms should also take into account the quality improvements required and the potential benefits consumers can gain by avoiding the expenditure on expensive bottled water and the environmental costs generated by the use of plastic.

1.7 Desalination

Several MENA countries depend partially on desalinated water for urban purposes. Among MENA countries, Gulf Cooperation Countries (GCC) are the most dependent on sea and brackish water desalination. In fact, Qatar and Kuwait rely totally on desalination, the United Arab Emirates relies 70% on desalinated water, and Saudi Arabia's reliance is nearly 60% (League of Arab States 2018). Figure 1.10 shows the annual volumes of desalinated water produced in some MENA countries. The UAE produced 1500 Mm^3 in 2012 and is ranked top in terms of volumes desalinated, followed by Saudi Arabia and Algeria. The countries that depend more on desalination are the rich-oil countries.

Desalination is an energy-intensive process. GCC countries have their own oil resources needed so far for desalination. Long-term solutions should be sought now to shift to renewable energy resources for desalination, to face the after-oil era.

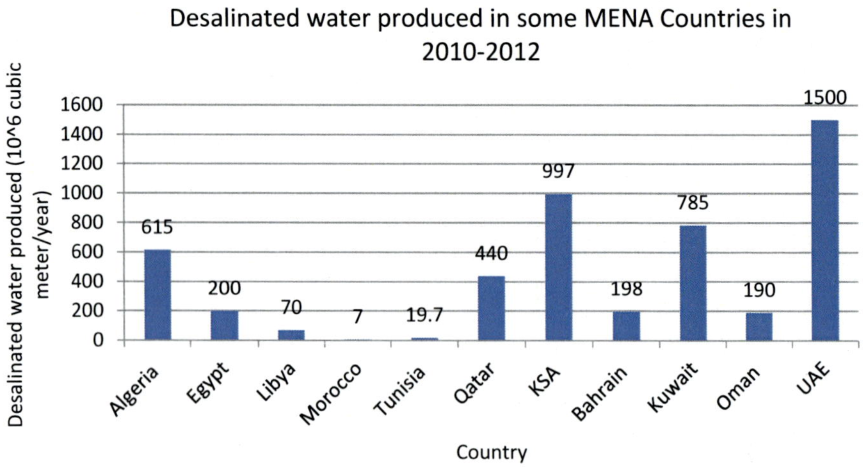

Fig. 1.10 Volumes of desalinated water produced in some MENA countries 2010–2012. (Source: Desalination Inventory Report for GCC Member States (2016) and FAO AQUASTAT database)

Brine disposal is another major concern. Most of the brine produced by Bahrain, Kuwait, Qatar, the UAE, Saudi Arabia, and Iran is dumped in the Arabian Sea causing seawater salinity increases and damages to the sea environment. According to Dawoud and Al Mulla (2012), seawater salinity in the Gulf is about 45 g/l, while in the vicinity of desalination plants seawater salinity is already varying from 50 to 55 g/l. GCC countries are now looking to establish a water network connection in order to desalinate more in Oman, in the open ocean where salinity is around 33 g/l, and transfer to the other Gulf countries (GCC 2016).

Most of desalination is undertaken by the private sector and foreign companies under different types of arrangements such as "Build, Operate and Transfer" and "Build, Own and Operate" and paid in hard currency. The contract type is "take-or-pay" where the buying entity agrees to buy a fixed volume of water year around regardless of fluctuation in demand. Usually desalinated plants are designed to supply for peak demand. The result is an excess of desalinated water during the winter months where demand is quite lower than summer. Zekri et al. (2019) proposed to inject the excess desalinated water, produced in winter, into aquifers to make use of the full capacity of desalination plants and enhance coastal cities' water security.

North African countries are increasingly using desalinated water for urban purposes but with much lower proportions. The trend in North Africa is toward a more intensive use of desalination to adapt to the population concentration in coastal areas, as well as to more recurrent droughts. Several coastal areas can no longer depend exclusively on water transfer from regions with excess conventional water due to higher water demand. Brackish water is also being desalinated in some interior areas of Algeria and Tunisia to improve the quality of delivered water. Oil-importing countries pay high costs for energy used for desalination and pay in hard currency. Furthermore, often contracts with foreign desalinating companies are paid in dollars. Given the devaluation occurring in several MENA countries, such as Egypt and Tunisia, the cost of desalinated water to the utilities is increasing dramatically making the goal of achieving cost recovery impossible since water prices do change only once a year in the best cases while the exchange rates are deteriorating on a daily basis. Thus, involvement of the local private sector should be considered in future desalination contracts in order to reduce the burden of exchange rates on water users and on financial stability of water utilities. Besides, the regions that depend on desalinated water should be charged more than the regions that benefit from conventional raw water. In most MENA countries, water prices are established at the national level and water users pay the same price regardless of the source/quality of the water delivered.

1.8 Funding

Most of the funding for the water sector is still coming from public funds (LAS 2018). This is fundamentally the result of low water tariffs in the urban sector as well as in the agricultural sector that makes investments by the private sector non-attractive. The required investments in the water sector for the coming decades are huge and estimated at US$ 100 billion per year, for the whole MENA region, in order to overcome water shortages in 2050 (Immerzeel et al. 2011). The funds vary from country to country and are needed for new investments as well as the replacement of old infrastructure. Although MENA has received an average $ 280 million per year during the period 2011–2013 ($ 0.50 per capita/year) in the form of development assistance for water resources (Jobbins et al. 2016) that represents only 0.3% of total required investments in the sector. The report by the World Bank (2018a) emphasizes that the participation of the private sector has been mainly focused on service efficiency so far. Some 28 million people living in MENA benefited from improved water services via public–private utility partnerships. The report stresses the fact that further participation of the private sector is conditioned by reforms in tariffs and subsidies and assurance of payments is addressed.

References

Abou Rayan, M. M., & Djebedjian, B. (2016). Urban water management challenges in developing countries: The Middle East and North Africa (MENA). In T. Younos & T. Parece (Eds.), *Sustainable water Management in Urban Environments. The handbook of environmental chemistry* (Vol. 47). Springer.

Bryden, J. M. (2017). Water, energy, and food in the Arab region: Challenges and opportunities, with special emphasis on renewable energy in food production. In K. Amer, Z. Adeel, B. Böer, & W. Saleh (Eds.), *Chapter 5. In the water, energy, and food security Nexus in the Arab region*. Cham: Springer International Publishing AG.

Chris, P., Steduto, P., & Karajeh, F. (2017). *Does improved irrigation technology save water? A review of the evidence*. Discussion paper on irrigation and sustainable water resources management in the near east and North Africa. FAO. 57 pages.

Dawoud, M. A., & Al Mulla, M. M. (2012). Environmental impacts of seawater desalination: Arabian gulf case study. *International Journal of Environment and Sustainability, 1*(3), 22–37. ISSN 1927-9566.

Desalination Inventory Report for GCC Member States. (2016). *Unified water sector strategy and implementation plan for the Gulf cooperation council of arab member states, years 2015–2035.*

Elmi, A. A. (2017). Food security in the Arab Gulf Cooperation Council States. In E. Lichtfouse (Ed.), *Sustainable agriculture reviews* (Vol. 25). Cham: Springer.

FAO AQUASTAT. (2018). *Food and Agriculture Organization of the United Nations, AQUASTAT database, data for 2008–2014*. http://www.fao.org/nr/water/aquastat/data/query/index.html;jsessionid=B213FE3569893C3BDBA96D00B7FE4878

GCC. (2016). *Unified water sector strategy and implementation plan for the Gulf cooperation council of arab member states years 2015-203*. 132 pages.

Immerzeel, W., Droogers, P., Terink, W., Hoogeveen, J., Hellegers, P., Bierkens, M., & van Beek, R. (2011). *Middle-East and Northern Africa water outlook*. Commissioned by the World Bank. World Bank task leader: Bekele Debele Negewo. FutureWater. 135 pages. http://siteresources.worldbank.org/INTMNAREGTOPWATRES/Resources/MNAWaterOutlook_to_2050.pdf

Jobbins, G., El Taraboulsi-McCarthy, S., & Florence, P. (2016, December). *Development finance for water resources. Trends in the Middle East and North Africa. Report*. Overseas Development Institute 2016. 31 pages.

Kotagama, H., Zekri, S., Al Harthi, R., & Boughanmi, H. (2016). Demand function estimate for residential water in Oman. *International Journal of Water Resources Development*. https://doi.org/10.1080/07900627.2016.1238342.

Kuper, M., Amichi, H., & Mayaux, P.-L. (2017). Groundwater use in North Africa as a cautionary tale for climate change adaptation. Water International, 42(6), 725–740. https://doi.org/10.1080/02508060.2017.1351058

LEA. League of Arab States. (2018). *Arab regional report: Pre-forum version*. World Water Forum 2018. Brazil.

ONAS. (2017). *Rapport d'Activite*. Office National de l'Assainissement. Tunisia. 34 pages. www.onas.nat.tn/Fr/telecharger.php?code=250

Sanchez, A., & Rylance, G. (2018). When the taps run dry: Water stress and social unrest revisited. *UNISCI Journal, 47*, 65–84.

Teodosiu, C., Gilca, A.-F., Barjoveanu, G., & Fiore, S. (2018). Emerging pollutants removal through advanced drinking water treatment: A review on processes and environmental performances assessment. *Journal of Cleaner Production, 197*(Part 1), 1210–1221. https://doi.org/10.1016/j.jclepro.2018.06.247.

Van Lavieren, H., Burt, J., Feary, D. A., Cavalcante, G., Marquis, E., Benedetti, L., Trick, C., Kjerfve, B., & Sale, P. F. (2011). *Managing the growing impacts of development on fragile coastal and marine ecosystems: Lessons from the Gulf. A policy report*. Hamilton: UNU-INWEH.

Whittington, D. (2016). Ancient instincts: Implications for water policy in the 21st century. *Water Economics and Policy, 2*(2) 1671002, 13 pages. https://doi.org/10.1142/S2382624X16710028

WITS. (2016). *World integrated trade solution data base*. https://wits.worldbank.org/CountryProfile/en/Country/DZA/Year/LTST/TradeFlow/Export/Partner/by-country/Product/16-24_FoodProd

World Bank. (2017a). *Tunisia WASH and poverty diagnostic. Phase 3 report: Synthesis and policy. Recommendations*. Report no: ACS17905.

World Bank. (2017b). Making the most of scarcity: Accountability for better water management results in the Middle East and North Africa (p. 270). https://doi.org/10.1596/978-0-8213-6925-8

World Bank. (2018a). *Beyond scarcity: Water Security in the Middle East and North Africa*. MENA development report. https://doi.org/10.1596/978-1-4648-1144-9

World Bank. (2018b). *World Bank database*. https://data.worldbank.org/indicator/ER.H2O.INTR. PC?view=chart

Zekri, S. (2010). Sustainable development for irrigated agriculture in Al-Hassa. Saudi Arabia. Report to FAO. Riyadh. Saudi Arabia.

Zekri, S., Al Harthi, S., Kotagama, H., & Bose, S. (2016). An estimate of the willingness to pay for treated wastewater for irrigation in Oman. Journal of Agricultural and Marine Sciences, 21(1), 57–63. http://dx.doi.org/10.24200/jams.vol21iss0pp57-64

Zekri, S., Triki, C., Al-Maktoumi, A., & Bazargan-Lari, M. (2019, January 22–24). *Optimal storage and recovery of surplus desalinated water*. 8th ICWRAE international conference on water resources and arid environments. King Saud University. Riyadh. Saudi Arabia.

Zheng, B., Zang, L., Li, W., Li, H., Wang, H., Zhang, M., & Song, X. (2018). Quantitative analysis of asbestos in drinking water and its migration in mice using fourier-transform infrared spectroscopy and inductively coupled plasma optical emission spectrometry. *Analytica Chimica Acta*. https://doi.org/10.1016/j.aca.2018.12.022.

Dr. Slim Zekri is Professor and Head of the Department of Natural Resource Economics at Sultan Qaboos University (SQU) in Oman. He earned his PhD in Agricultural Economics and Quantitative Methods from the University of Cordoba, Spain. He is Associate Editor of the journal *Water Economics and Policy*. He has worked as a Consultant for a range of national and international agencies on natural resource economics, policy and governance, agriculture, and water economics in the Middle East and North Africa. He is Member of the Scientific Advisory Group of the FAO's Globally Interesting Agricultural Heritage Systems. His main research interests are water economics and environmental economics. In 2017, he was awarded the Research and Innovation Award in Water Science from the Sultan Qaboos Center for Culture and Science.

Ms. Aaisha Al Maamari obtained her master's degree in Agricultural Economics from Colorado State University, USA, in 2017. In 2018, she joined as a Lecturer in the Department of Natural Resource Economics, Sultan Qaboos University, Oman. Her research interests are on international trade, development economics, and water policy.

Chapter 2
Water Policy in Algeria

Nadjib Drouiche, Rafika Khacheba, and Richa Soni

Abstract Water scarcity is a reality in Algeria. However, this stark fact is sometimes misunderstood. High population growth coupled with industrialization calls for a sustainable water use pattern in industrial, agricultural, and domestic sectors. The problems caused by water scarcity imply important changes in the criteria and objectives of water policies. The major water issues in Algeria can be attributed to both policy implementation failure and a lack of on-the-ground application of regulations. For a better tomorrow, it is pressing to integrate policy frameworks with particular attention to efficient utilization, rationalization, and conservation. The proposed approach is illustrated in the paper by the case study of water policy analysis for Algeria.

Keywords Algeria · Water policies · Institutions · Pricing · Irrigation schemes

2.1 Introduction

Algeria is located in North Africa. Most of the country is covered by the Sahara Desert. Then it becomes evident that the country is a water-scarce nation. The Algerian population is expected to reach 45 million by 2020. Not only is the country affected by the severity of its climate and geography but also by the overexploitation of its water resources to an alarming extent. The quality of water management and

---

N. Drouiche (✉)
Centre de Recherche en Technologie des Semi-conducteurs pour l'Energétique (CRTSE), Algiers, Algeria

R. Khacheba
Ecole Nationale Supérieure d'Agronomie Kasdi Merbah, El Harrach, Algeria

R. Soni
School of Engineering, Indian Institute of Technology, Mandi, Suran, India

© Springer Nature Switzerland AG 2020
S. Zekri (ed.), *Water Policies in MENA Countries*, Global Issues in Water Policy 23,
https://doi.org/10.1007/978-3-030-29274-4_2

service is one of the key priorities of the government as various forecasts predict that climate change could severely affect Algeria in the form of more frequent drought and flood cycles (CEDARE 2014; Drouiche et al. 2012; Hammouche 2011).

In Algeria, the coastal areas have a mild climate with hot summers and cool and rainy winters. In the highlands, summers are hot and dry. Winter rains in the highlands begin in October. However, most of the territory is occupied by the Sahara, which explains the distribution of population density. Ninety percent of the population is living along the coastlines (Agence du Bassin Hydrographique Constantinois-Seybousse-Mellegue 2009; Hammouche 2011).

Recent research shows that the water resources in Algeria are estimated at 17 billion m^3, with surface water estimated at 10 billion m^3; groundwater at 7 billion m^3 mainly in the Sahara. The aquifers situated in the north are exploited to 90%, with 1.9 billion m^3 per year with some aquifers being overexploited. In the Sahara region, the extracted volume of groundwater is around 1.7 billion m^3 per year (Drought management strategy in Algeria 2014).

The weather information services on water management are delivered through a National Agency for Water Resources via 220 hydrometric stations, 860 rainfall stations, 36 rain gauges, and 56 full weather stations (Drought management strategy in Algeria 2014). To support a growing and rapidly urbanizing population, Algiers hopes to employ technological solutions to maximize the country's water supplies. The emergency program dealing with the crisis and the disruption of water supply has highlighted the random nature of water resources and recommended to relying more heavily on nonconventional water resources such as seawater desalination and reuse of treated wastewater (Ministère des Ressources en Eau 2011), leading to the implementation of a new sector policy of water resources (Boye 2008; Drouiche et al. 2011; Mooij 2007).

2.2 Legal and Organizational Aspects of the Water Sector

The ministry of water resources (MRE) is the most important player in the Algerian water sector. Its mandate is given by Decree no 16-89 of 2016, having the responsibility of elaboration and implementation of policies and strategies in the context of water resources and environment protection. MRE controls both the water and environment sectors. Several public administrations and organizations are subordinated to it (CEDARE 2014; Kettab 2001; The World Bank Report No.: 36270 – DZ 2007). The Algerian water sector is characterized by few companies/utilities. Figure 2.1 lists the agencies and their prime responsibilities (Drouiche et al. 2012; Ministère des Ressources en Eau 2011).

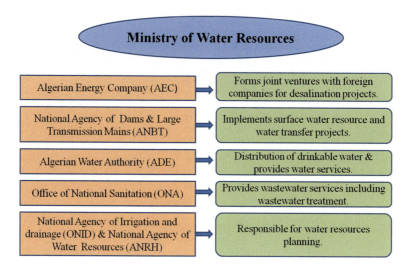

Fig. 2.1 Agencies of water sector and their roles

2.3 Agricultural Water

2.3.1 Surface Water

There are 17 major watersheds in Algeria. The country receives rainfall in an annual average of 89 mm, which allows a flow of 211 billion m^3. Rainfall is variable across the country with 350 mm average annual rainfall in the west and as much as 1000 mm in the north east. Rainfall decreases rapidly south of the Saharan Atlas range and toward the Sahara Desert. The surface water is 10 billion m^3 annually, distributed according to five watersheds, as shown in Fig. 2.2. Most of the surface water resources are concentrated in the North, along with the majority of the population (Groundwater Management in Algeria. Draft Synthesis Report 2009; Knoema 2014).

Only the rivers in the northern coastal region are perennial, flowing all year round. In the south, wadis (ephemeral rivers) drain to closed internal sinks—chotts or sebhkas—which are subject to high evaporation rates. In the drier area to the north of the Atlas, soils are generally suitable for agriculture, but water availability is a key constraint. Withdrawals for agriculture are estimated to be 3940 Mm^3/year, which means that some of these surface waters are being fully used mainly due to expansion of the irrigated land of agriculture irrigation (CEDARE 2014).

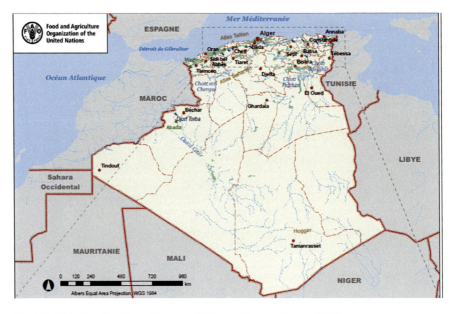

Fig. 2.2 Major Surface Water Features of Algeria. (Source: fao.org. 2015)

2.3.2 Groundwater

Groundwater is the major source of drinking water and its use for irrigation is forecasted to increase substantially to combat the growing food insecurity. Because of its geographical location, Algeria undergoes the influence of two climate types: the Mediterranean type in the north and the Saharan type in the south. Groundwater withdrawals are roughly double the annual recharge rate. The total available renewable potential is estimated at about 2.7 billion m^3/year in the northern Atlas region and 5 billion m^3 in the southern Saharan region. These aquifers are fed essentially by rainfall, whose distribution is irregular both in time and space. Agriculture is the country's primary user of total water, taking almost 4000 MCM/year and is the sector having the highest impact on aquifers, causing a threat for groundwater contamination (Ghodbane et al. 2016; The world bank Report No.: 36270 - DZ 2007).

2.3.2.1 Fossil Water vs Renewable Groundwater

Water is usually a renewable resource, but in some cases, it can be fossil. The fossil water requires careful exploitation and usage, as it is a nonrenewable resource. The largest fossil groundwater volumes are found in the large sedimentary aquifers in the Sahara where the climate is harsh desert with little or no rainfall.

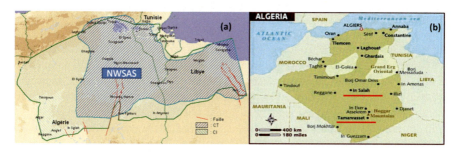

Fig. 2.3 (a) Hydrogeological sketch map of the North-West Sahara Aquifer System (Foster and Loucks 2006). (b) Map showing position of Tamanrasset

Table 2.1 Projections of the irrigated surfaces evolution in the NWSAS region

Country name	Surface (ha) in 2000	Surface (ha) in 2020	Surface (ha) in 2050
Projections of the irrigated surfaces evolution in the NWSAS region			
Algeria	170,000	300,000	340,000
Demographic Projections (number of capita of the NWSAS region)			
Algeria	2,600,000	3,700,000	4,800,000

Algeria manages and shares its deep fossil groundwater (i.e., The Continental Interlayer fossil aquifers) with Libya and Tunisia as a member of the North-Western Sahara Aquifer System (NWSAS). The agreement is subjected to an international cooperation setup in 2008 between the three countries to sustainably manage the groundwater resource. The latter involves the sharing of two major confined fossil aquifers—the Continental Interlayer sandstone aquifer and the shallower sandstone aquifer known as the Terminal Complex. This system, as shown in Fig. 2.3a, covers an area of more than 1 million km^2, 69% of which is in Algeria, almost 8% in Tunisia, and 23% in Libya. The projections of the increasing pressure on NWSAS water resources for the few coming decades are alarming and are presented in Table 2.1.

This resource is largely nonrenewable, not fully exploitable but confronted to risks due to exploitation, including water salinity and falling water table level, which seriously threaten the sustainability of the economic development in the region.

Algeria started pumping fossil water reserves through a large-scale transfer project to transfer water from Ain Salah to Tamanrasset, over a distance of 750 km, completed in 2011. This project supplies potable water to the city of Tamanrasset to balance the demand. The Ain Salah-Tamanrasset water transfer canal has a conveyance capacity of 100,000 m^3/day. This attitude toward abstraction is prevalent across the entire region (CEDARE 2014; Edmunds et al. 2003; Foster and Loucks 2006; Groundwater Management in the Near East Region Synthesis Report 2011; Sekkoum et al. 2012).

2.3.2.2 Aquifer Depletion

In a semiarid region like Algeria, where groundwater is a primary source of water, intensive irrigation may threaten future water security. In addition, with anticipated shifts in precipitation patterns induced by climate change, groundwater's value as a strategic reserve is increasing in the country. While farmers in Algeria have significantly improved their livelihoods and household food security, aquifer depletion and groundwater pollution are also a direct result of this intensive use of groundwater for irrigation purposes. In the North-Western Sahara, where people are dependent on the North-Western Sahara Aquifer System, depletion of aquifer reserves is taking place with a high dependency on irrigation for agriculture. During the last 30 years, 0.6–2.5 billion m^3/year of water has been abstracted from this system. Algeria is planning to increase the extraction as a response to climate change.

In the Sahara, the *foggara* has played a leading role in the field of abstraction of groundwater distribution and sharing through formal and strict rules. Foggaras are tilted underground tunnels draining water from the water reserve to the field of irrigation. The water flows downstream owing to gravity. It consists of a series of wells 3–12 m apart and a tunnel 50–80 cm wide and 90–150 cm high. Fogarras are still functional and are used to irrigate oases like Touat and Tidikelt group and Gourara oases.

In the northern part of the country, groundwater resources are still available, either in the form of springs or shallow wells. Although significant progress has been made in groundwater governance, the overexploitation due to irrigation over the past two decades has particularly affected the northern areas of the country. Local estimates suggest that by 2025 groundwater supplies could be fully exploited and in some places, they will be overexploited as they are already in parts of the north (Maliva and Missimer 2012; Remini and Achour 2013; Good Practices in Agricultural Water Management Case Studies from Farmers Worldwide 2005).

2.3.3 Property Rights

Algeria has adopted a participatory process to review its legislation and policies and improve its legal and policy framework. Under the 2005 Article 3 of the Water Law, the right to access to water and sanitation to satisfy the basic needs of the population, respecting equity is recognized (Export.gov 2019; Investment Climate Statement – Algeria 2015; Righttowater.info 2019).

Water rights are private in some parts of the country. Thus, in some regions of Algeria partnership agreements exist between the owners of unirrigated land and the owners of water quotas. The share of water received by each recipient is determined based on the size or level of investment contribution. "Association sharing" exists in a region of Southern Algeria between owners of unirrigated land and owners of water quotas for the production of palm dates. In such an agreement the owner of the land transfers half of his property rights to the owner of the water quota. After few years, when the tree bears fruits, the "association" comes to an end and the land owner gets permanent ownership of the water on his land. Both participants benefit through such association as each gets half the ownership of palm groves (Benmehaia and Brabez 2016; Benmehiaia and Brabez 2016; Laoubi and Yamao 2008).

2.3.4 Reuse of Treated Wastewater

During the last three decades, Algeria has suffered from recurrent progression of periods of drought (Intergovernmental Panel on Climate Change (IPCC) 2007). A decrease in annual rainfall of between 10% and 20% was observed in the western region of the country (Meddi and Meddi 2007). The use of groundwater in the northern region has already reached its limits leading to higher pumping rates that have affected the groundwater levels in this region. Significant rise in the population, agricultural irrigation, and economic development have exerted further strain on the water resources. Climate change has also intensified the water scarcity issue. Therefore, reuse of wastewater has become a necessity as properly treated wastewater has been successfully employed to meet nonpotable water needs.

Wastewater reuse alternatives can be generally listed as agricultural reuse, urban and landscape reuse, industrial reuse, domestic reuse, and groundwater recharge. The possibility of reusing treated wastewater for the above-mentioned alternatives depends upon several factors, which include water quality requirement of each user, location of user, probable health risks, government regulations, and cost requirement. Owing to the growing population, water requirement by the agricultural and industrial sector is increasing. As a result, wastewater reuse contributes toward the sustainable management of water resources. The volume of sewage water in Algeria is estimated to be 600 million m^3/year. This number is expected to reach 1.2 billion m^3 in the year 2020 (ONA), 2011, http://www.ona-dz.org/ (accessed 06.06.12).]. Currently, the number of wastewater treatment plants in Algeria exceeds 60, with a total treated volume of 1 Mm^3/day approximately. The Government investments in this subsector over the last 30 years amounted to USD 15 billion (Drouiche et al. 2012). A list of treatment plants is shown in Table 2.2.

Table 2.2 List of major treatment plants, their capacity and volume they treat

Name	Wilaya	Year of service	Capacity (m^3/day)	Treated volume
BBA	BBA	2008	2500	30,000
Ibn Ziad	Constantine	2009	5000	69,120
Ain Hout	Tlemcen	2009	9300	30,000
Ghriss	Mascara	2012	1000	3,700
Baraki	Alger	2013	76712	150,000
Annaba	Annaba	2013	83620	116,000

2.3.5 Food Safety/Vegetables/Heavy Metals

The agriculture sector in Algeria uses 70% of the water resources and is most exposed to weather and therefore most sensitive to drought. In Algeria, treated wastewater is used for unrestricted irrigation, industrial use, and restricted agricultural use according to the degree of treatment. Treated wastewater contains significant amounts of nutrients (nitrogen, phosphorus, and potassium) ranging from 20 to 60 mg/L nitrogen, 6–15 mg/L phosphorus, and 10 to 40 mg/L for potassium (da Fonseca et al. 2005). These nutrients are essential to crops for high yields and undesirable for environmentalists in aquatic compartments, whose excess presence is detrimental to the environment and public health (Mohammad and Mazahreh 2003). Although much progress has been made on laws and standards for wastewater reuse, the critical water scenario suggests the need for further development of wastewater reuse standards and related laws (Intergovernmental Panel on Climate Change (IPCC) 2007; Hamdy and Lacirignola 2005; Meddi and Meddi 2007).

2.3.6 Pricing

2.3.6.1 Cost Recovery/Maintenance of the Irrigation Schemes

Water pricing and cost recovery of irrigation investments, operation, and maintenance have been contentious issues for many decades. In Alegria, water pricing is considered as a significant economic tool in the reform process of water demand management. Over the two last decades, the country invested in irrigation to secure and increase water supply in order to develop the economic sector, improve food security, and target populations in less favored rural areas. In this regard, reforms in water resources management have been established to improve the performance of irrigation schemes, such as the national plan of agricultural development (PNDA) in 2000 and the water pricing policy of 2005. Other steps taken to improve cost recovery include improved irrigation services and more transparent decisionmaking. The success of these steps relies on government policies as well as institutional arrangements, including the basic legal system. The agricultural price policies have a negative impact on farmer's incomes, which affect policies of water pricing or need complementary policies addressing the issue of acceptance to the farmers. Electricity pricing in the agriculture sector is expected to be the most efficient way to begin managing and regulating the use of groundwater by farmers (Algeria MWR et al. (2014) (Table 2.3).

Table 2.3 Electricity consumption by agriculture (2012) in ktoe [33]

Sector	In ton of oil equivalent (TEP)	In %
Agriculture	1,069,935	3

2.3.6.2 Irrigation Organizations and Farmers' Participation

The National Office of Irrigation and Drainage (ONID) allocates volumes of water to the irrigation schemes according to the needs expressed by farmers. The process begins at the starting of irrigation season. The user submits his water demand to the agency, by specifying the number of hectares, the crop type, and volume of water desired. The ONID forwards the volume requested to the DHA Directorate before the start of the irrigation season. The request is subsequently processed in the Sub-directorate of the Operation and Regulation of Agricultural Water Management (SDERHA) (Laoubi and Yamao 2008). The final decision on allocations is determined according to the available storage in the dams, after subtracting the volume of water allocated to urban use. The MADR informs the DSA (Directorate of Agricultural Utilities of the Provinces), which, in turn, informs farmers (Chamber of Agriculture) of quota availability. Farmers contact the ONID and organize themselves for the water release.

Crop irrigation accounts for approximately 70% of the total water consumption and covers an area of 865,286 ha (i.e., 2% of the agricultural area), with a predominance of medium and small-scale schemes (88% of the area). These systems were partially or entirely created by farmers using shallow wells, deep tube wells, small reservoirs, wadis (spate irrigation systems), springs, and ghotts (small oases in the South).

In Algeria, there is little research on farmers' cooperatives and organizations. The main studies showed that farmers' cooperation behavior relies on their perception of professional cooperatives and that the educational level is an important factor critical to farmers' participation in cooperatives and associations. They concluded that the government should increase its efforts in promoting and publicizing the benefits of cooperation in more effective ways (Benmehaia and Brabez 2016; Benmehiaia and Brabez 2016).

2.3.7 Irrigation Efficiency

2.3.7.1 Technology Adoption/Subsidy and Other Policies

Many challenges that re already confronting irrigation development in Algeria and that will become steadily more acute as population grows and climate change adds stresses on the available freshwater resources. Careful management of scarce water resources is thus essential to improve food security (Dinar and Mody 2004). Since the 1980s, the government of Algeria began development of the National Irrigation Policy, which is designed to fully address the challenges that the irrigation sector faces and promote effective irrigation development. The concept of integrated water resources management was introduced by the Algerian water policymakers. It aims at providing a vision and step-wise prioritization of irrigation development in the country. Several reforms have been tried since, including water conservation, good irrigation development and management practices and use, improving water use efficiency, and the sustainability of irrigation schemes. The principle of cost recovery has been established, and the irrigation water prices increased in 1998 and 2005. A number of innovative technologies (drip, sprinkler) are financed through water sales and state subsidies have been tested and adopted in Algeria (Dinar and Mody 2004; Easter and Liu 2005; Ghazouani et al. 2012). Over the last decades there has been a major change in the irrigation technology used in Algeria. There has been a general move from traditional flood irrigation to application of more efficient irrigation technologies such as central pivots, drip, and microsprinklers.

2.3.7.2 State-of-the-Art Technology/Smart Irrigation & Innovations

Facing severe water shortages, Algeria is trying hard to save irrigation water by investing in water saving technologies and changing cropping patterns toward less-water-demanding crops. Cereal crops such as wheat and barley are grown along coastal areas and in some of the mountain valleys where rainfall is plentiful. Potato is also grown. However, only around 3% of Algeria's land is suitable for arable farming. The slopes of Algeria's northern mountains and plateaus are used for pastoral farming, mainly sheep, cattle, and goats. Further south and across the desert regions, date palms are common in oases. Poor irrigation management has often resulted in several sites to soil salinization and groundwater contamination and pollution. New irrigation technologies and decision support tools are continually being innovated in Algeria and worldwide. Water use efficiency and energy use efficiency are the main focuses of these innovations. Fortunately, efficiency is linked to better quality production and improved profitability.

2.4 Research

New technologies, mainly seawater desalination and water reuse processes, which address the emerging challenges, are easily embraced by the water sector to increase water availability. Some cooperation agreements exist between the ADE and ONA and research institutes for water-related issues. Algeria and the European Union (EU) signed the Prima agreement on scientific cooperation to increase research in the key sectors of water and agriculture. The PRIMA initiative (2018–2028) aims to develop new solutions for sustainable water management and food production (PRIMA; https://www.euneighbours.eu/en).

FAWIRA project is launched in Algeria with the objective of strengthening the cooperation capacities of Algeria's "National Institute of Agronomic Research"— INRAA in the context of the European Research Area and development to the Food, Agriculture and Water center of excellence. This is in turn facilitates its participation in European and regional collaborative research initiatives. To ensure adequate coverage of research areas, INRAA deploys an intensive cooperation both nationally and internationally. The international project portfolio consists of several projects dedicated to the improvement of wheat, building research capacity, renewable energy, the fight against desertification and rural development in fragile ecosystems (mountains, plains, and Sahara areas), food security, the application of nuclear techniques to agricultural areas (Pathology and Animal Health, irrigation, crop breeding, bioclimatology), mobilization of water resources, and soil conservation.

The Centre for Development of Renewable Energies (CDER) was established as a research center specialized in renewable energies, resulting from the restructuring of the Algerian High Commission for Research. The center is responsible for developing and implementing programs of research and development in the field of science and technology, energy systems exploiting solar energy, wind, geothermal, and biomass.

2.5 Food Security Versus Virtual Water/Food Imports

By importing food and agricultural goods, Algeria copes with the heterogeneous global water distribution and often relies on water resources available globally. The marked decrease of import observed in Algeria is likely due to the expansion of the irrigated areas in the last two decades. However, at the country level, there is an extreme variability in terms of total renewable water resources (TRWR), the TRWR per capita is less than 600 m³/capita/year, the threshold that corresponds to the water scarcity levels proposed by Falkenmark (1986). As a result, the country imports most of its basic foods and all of its livestock feed, one-half of the public expenditure in the budget of the Ministry of Agriculture is allocated to financing the price support program for wheat.

Cereals represent 38% of Algeria's total food import bill (2015) and also the top food import. Algeria imported 6–7 MMT per annum of total wheat over the past 5 years of which bread wheat always represents 75–83% of the wheat imports. Algeria imported an average of 222,000 MT of pulses per year over the past 5 years mostly from Canada, China, Mexico, Argentina, and Turkey. Barley imports increased following an average crop production in 2013/2014, and as the production was revised downward in 2014/2015, more barley was imported to meet the demand for animal feed.

2.6 Water Salinity/Other Pollution Problems

Algeria faces severe water scarcity and hence resorts to groundwater to cover water demand. Overexploitation of groundwater, however, often causes intrusion of seawater into coastal fresh water aquifers, and as a result there is fresh water shortage. Often, emergency measures are being imposed to counter fresh water shortages. Water scarcity is further complicated by lack of sewage control and pollutants from the oil industry and other industrial effluents.

Nitrates are a key ingredient of manures used in Algeria as they serve as a nitrogen source for plants. Part of these manures is absorbed by plants while the rest is collected in surface water bodies such as the Dam of Beni Haroun in Algeria or in the groundwater (Cheurfi et al. 2009). Nitrate pollution is hazardous as it is toxic for humans if ingested in too large quantities. Also, it participates in eutrophication along with phosphates.

2.7 Urban Water

2.7.1 Sources of Supply

Algeria undertook over the last decades a vast program of rehabilitation and extension of the city networks of drinking water distribution and treatment. The reform also touched on the capacity building of management of the water public service. The rate of connection of the population to the public network of urban water went from 78% in 1999 to 94% in 2011, with an average consumption of 170 L/capita/day. In Algiers, since 2006, urban water supply is managed by a public company called SEAAL in cooperation with a private partner, SUEZ environment. The company provides drinking water services to approximately 3 million people. In 2011, the Algerian authorities renewed and extended the contract with SUEZ for 5 years to help modernize the water and wastewater management services for Algiers. The contract covers the provinces of Algiers and Tipaza and serves 3.8 million citizens approximately.

In Oran, the water utility "Société de l'Eau et de l'Assainissement d'Oran" (SEOR) was established in 2008. It is a 100% state-owned company, with sufficient funding (a 15 billion Euro investment over 5 years) and operated by a private company, Agbar, through which the company provides drinking water to over 1.6 million people. The access to water 24 h a day has increased from 10% to 99.3% in just over 5 years.

In June 2008, the Société des Eaux de Marseille (SEM) of France, a subsidiary of Suez and Veolia, was awarded a Euro 28 million and 5.5-year management contract for Constantine. The objectives of the management contract are to provide good quality water on a continuous basis, to improve bathing water quality on the beaches, to rehabilitate infrastructure, and to improve customer satisfaction. The operator began its work in 2008. The contractual partner is the water and sanitation utility for Constantine, SEACO.

Other cities in Algeria rely on the state-owned company Algerienne des eaux, which continues to provide water and sanitation services without partnerships with the private sector. Eighty percent of water distribution systems in Algeria are under its responsibility. Since 2008, desalinated water has been a non-negligible part of the water supply in the Northern part of the country. Algeria has 15 seawater desalination plants along its coast in 2011 with a capacity of 2.3 million m^3/day. Growing desalination capacity has helped increase water supply in the coastal cities by about 30%. The Magtaa plant, which began operations in 2014, has a capacity of 0.5 million m^3/day, provides 5 million people with drinking water for the eastern population. There are plans to expand desalination capacity for seawater and brackish groundwater in the near future. Besides these sources of water supply and management, the authorities launched a series of major projects to ensure water supply. Among these measures, the 2015–2019 five-year plan plans to spend 262 billion USD in developing new basic infrastructure and completion of ongoing projects.

2.7.2 Regulations, Priority, and Allocation

In the year 2000, the Ministry of Water Resources was created under Executive Decree no. 2000–324. All directorates related to irrigation were transferred from the Ministry of Agriculture to the Ministry of Water Resources. The latter launched broad institutional reforms. Five public agencies have been established that are in charge of developing infrastructure and managing water services, sanitation, and irrigation (Drouiche et al. 2012).

In the year 2005, a new water law was set up (Law No. 05-12, 2005) that aims to establish the principles and rules for use, management, and sustainable development of water resources as an asset of the national community. It stressed upon the rational use of water. Executive Decree no. 07-149 of 20 May 2007 established detailed rules for the use of treated wastewater for irrigation as well as standard specifications relating thereto. Executive Decree no. 07-270 of 11 September 2007 established the conditions and procedures for establishing a pricing system for water for irrigation. Executive Decree no. 07-399 of 23 December 2007 relates to the qualitative protection of water resources. Executive Decree no. 07-69 of 19 February 2007 established the conditions and procedures for authorizing the grant of use of thermal waters (Ministère des Ressources en Eau 2011).

2.7.3 Drought Management

Algeria has been experiencing more frequent drought events over the last two decades. Droughts that hit most of the regions of North Algeria in the early 1980s and the early 1990s had substantial negative effects on agricultural production, natural resources, and socioeconomic aspects. Review of literature on drought in Algeria indicates that drought occurs especially in the dry south Oran region (Hirche et al. 2011). In Algeria, average annual rainfall has decreased by over 30% in recent decades. The 2001 drought in Algeria was caused by a low precipitation rate in all the territory. The annual average precipitation has been lower than the minimum measured in the historical series from 1991 to 2015. This extreme reduction of rainfall resulted in significant impacts on water stored in reservoirs, potable water availability, and water quality. This situation called for the implementation of special emergency plans and eventual drought, implementing management measures such as irrigation restrictions and setting emergency measures. Grain production in Algeria fell by 11% to 3.3 million tons in 2016–2017 from 4 million tons in 2015–2016.

2.7.4 Water Transfers and Water Markets

The water transfer process was a top priority for the Algerian Government not only for immediate drought response, but also as part of a comprehensive, long-term water management policy, as outlined in the national water strategy. Improvements to the water transfer process and creation of water markets would play an important role in implementing a comprehensive long-term water management policy. Achieving this goal requires the use of large transfers, and to rely on nonconventional resources, especially seawater desalination and reuse of treated wastewater, and water saving.

The Taksebt Dam supplies drinking water to a population of 4 million people. Mostaganem–Arzew–Oran (MAO) project transfers an annual volume of 155 million m^3 of water in the east of the country. The Setif-Hodna-El Eulma transfer project is comprised of west and east system transfers. The west system enables the transfer of an annual volume of 122 million m^3 of water to the 550,000 inhabitants of Setif and the irrigation of 13,000 ha of plains in Setif. The east system allows the transfer of an annual volume of 190 million m^3 of water to the 700,000 inhabitants of El Eulma city and the irrigation of 30,000 ha of land (Ministère des Ressources en Eau 2011). Beni Haroun dam is intended to transfer the water through pumps to six provinces, namely, Batna, Khenchela, Mila, Oum El Bouaghi, Constantine, and Jijel–Mila region (Fig. 2.4).

2.7.5 Water Quality

In Algeria, the quality of water is affected by various forms of pollution. Water quality, neglected for a long time, has now become the focus of attention. Data indicate that most water resources are polluted, and major pollution sources are municipal wastewater discharge, industrial effluents, and agricultural activities. The estimated volume of wastewater generated from urban areas is around 1 billion m^3/year (Karrou et al. 2011). Despite major efforts during the last decade to build wastewater facilities, the overall quality of water remains at risk and could worsen from industrial, domestic, and irrigation expansion. Salinity is also a major constraint. Salinity levels of surface water vary between 0.8 and 1.5 g/L with the majority of resources having a salinity of less than 1 g/L. The majority of groundwater resources are of poor quality due to high salinity; in the north, most groundwater is nonsaline

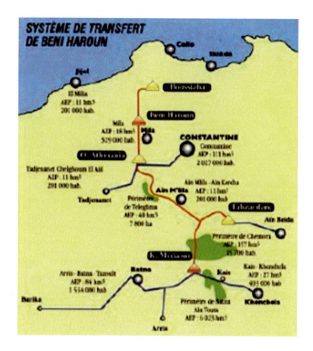

Fig. 2.4 Map of Beni Haroun transfer

with less than 1 g/L salinity. In the south, the salinity levels are variable and some sources have high salinity levels of up to 8 g/L. Although most of the domestic wastewater collected is treated, untreated sewage is still being discharged into natural water bodies. Industries discharge untreated effluents into natural water bodies in violation of government regulations. About 200 million m^3/year of untreated industrial wastewater is discharged into the environment. Uncontrolled and improperly monitored leaching practices and agricultural drainage that includes nitrates and phosphorus from fertilizers pollute water. Pesticide residues can also be detected in some surface water. Leaching from improperly discharged and untreated solid waste also pollutes water. Agricultural drains also receive domestic pollution.

2.7.6 Desalination

Over the two last decades, Algeria has experienced a dramatic demographic shift as large numbers of rural dwellers have moved to cities. To alleviate the water shortage, Algiers needed to find a sustainable, long-term water supply that could meet the expanding urban water demand. Investment in secure sources of water like seawater desalination enabled rapid and affordable solutions without adverse environmental impact. Such supplies are always available in times of drought and scarcity and ensure response to demand growth, climate change, and increasing weather volatility (Drouiche et al. 2011). To this end, the Algerian government had embarked on a

Table 2.4 Seawater desalination capacity in Algeria

Project	Capacity (m³/h)	Commissioning	Partners
Kahrama	86.880	Since 2006	J. Burrow Ltd:50%
Hamma	200.000	Since 2008	GE Ionics "Etas-Unis":70%
Skikda	100.000	Since 2009	Geida (Befesa/Sadyt) "Espagne": 51%
Beni Saf	200.000	Since 2010	Obra/Espagne: 51%
Souk Tlata	200.000	Since 2011	TDIC (Hyflux/Malakoff) "Singapour:51%
Fouka	120.000	Since 2011	AWI (Snc Lavalin/Acciona): 51%
Mostaganem	200.000	Since 2011	Inima/Aqualia "Espagne" :51%
Honaine	200.000	Since 2012	Geida (Befesa/Sadyt) "Espagne" :51%
Cap Djinet	100.000	Since 2014	Inima/Aqualia "Espagne" :51%
Tenes	200.000	Since 2015	Befesa "Espagne" :51%
Magtaa	500.000	November 2015	Hyflux "Singapour": 47% ADE, 10%

major and large-scale program of investment in seawater desalination to meet the demand. Oran was the first major city to use desalinated water for drinking water supply through Kahrama project, which is an independent water and power project (IWPP). The facility produces a net power output of 300 MW coupled with an average annual water production of 80,000 m³/day, and by 2008 Algiers was the second city to rely on desalinated seawater. Nine other large-scale SWRO plants are in operation at several locations for the purpose of supplying drinking water (Algerian Energy Company (AEC) 2011; Bessenasse et al. 2010; Ghaffour 2009; Kettab 2001; Water Desalination Report 2011; Terra 2011). Desalination plants in Algeria have a cumulative capacity of about 2 million m³/day (see Table 2.4). They are in the form of public and private partnership. The public partner is the Algerian Energy Company that is a consortium between the National Oil Company (sonatrach) and the National Company of Electricity, Sonelgaz.

Seawater desalination in Algeria is becoming a practical technology even though the prices are still higher than conventional water. This is a result of water scarcity as a result of climate change and concentration of population in coastal cities.

2.7.7 Wastewater Treatment

The average production of wastewater per capita per day is estimated at around 140 L/day. The actual total volume of sewage discharged in Algeria is estimated at about 731 hm³/year. Sixty-three cities are served by 61 large wastewater treatment plants. The treatment provided is mainly secondary (56%) and primary treatment (44%). Disposal of treated (510,000 m³/day) as well as untreated sewage (290,000 m³/day) is discharged into the rivers or to the sea. A limited quantity of treated sewage (4.5%) is reused to irrigate about 7500 hectares.

Municipal sanitary services are carried out by a set of authorities affiliated to the ministry of water resources (MRE). Under MRE, the national office of Sanitation

(ONA) manages 68 wastewater treatment plants (WWTP) and 41,000 km network. ONA is responsible for operating municipal wastewater treatment plants and sewage collection systems throughout Algeria. The ONA has embarked on an approach to environmental management according to ISO international standard 14,001 with 2004 version. With the achievement of approximately 1500 km of collectors per year, the national total sanitation network is expected to reach 54,000 km by 2020. Sewage collection systems are delegated by a management contract established between ONA and ADE on one side and foreign private operators (SUEZ environment in Algiers, Eaux de Marseille in Constantine, Agbar in Oran) on the other side under the oath of the Ministry of Water Resources to improve sewerage collection and to connect urban households to sewers (Cheurfi et al. 2009; Kettab 2001; Office Nationale de l'Assainissement (ONA) 2011; Terra 2011).

Overall, the National Park of sewage treatment plants has 134 sewage treatment plants, 61 sewage treatment plants (activated sludge), and 73 lagoons (stabilization pond, aerated lagoon, sand filter, reed bed sewage treatment, garden filter) with a total capacity of 800 million m^3/year in 2010 for sewage only. The WWTP and lagoons have laboratories for daily control of water quality at the inlet and outlet works and the quality of sludge (DB05, COD, TSS). These laboratories are backed by the central laboratory of the ONA, which provides more analysis of heavy metals on water and sludge treatment plants. In addition, a project to implement a Geographic Information System on a national scale is currently being developed (Kettab 2001; Office Nationale de l'Assainissement (ONA) 2011; Terra 2011).

The program for 2016–2020, it is projected the achievement of 50WWTP, which would make reach the treatment capacity in 2020 to 1.2 billion m^3.

2.7.8 Wastewater Reuse in the Urban Sector

To combat the water scarcity the treated wastewater is reused for different purposes, namely: in the preservation of the environment and water resources, industrial use, including the cooling of industrial installations, and artificial recharge of aquifers.

2.7.9 Urban Water Pricing

Responsibility for drinking and wastewater operations in the major cities of Algeria was transferred through contracts involving multinational corporations. These corporations became increasingly involved in the provision of urban water infrastructure, in investment, management, and sewerage services. Algeria Water Company (ADE), a public utility, is responsible for the management of these services in the remaining cities. Price rates are often determined by considerations of political acceptance and are therefore subsidized. All households connected to the water network in Algeria benefit from subsidies for their water consumption. The authorities

Table 2.5 Water prices and subsidies in Algeria 2005

	Price in Algerian Dinars	Price in USD	Percentage of subsidies
Average	31	0.30	57
Block 1 0–25 m^3	28.3	0.27	60
Block 2 26–55 m^3	24.9	0.24	65
Block 3 56–82 m^3	31.6	0.30	56
Block 4 > 83 m^3	46.7	0.44	35

that set prices for drinking water must steer a course between political acceptability and the need to cover cash deficits. The water price is stipulated by Law no. 05-12 of 4 August 2005. The prices are volumetric with four blocks. Billing is quarterly. The average price is $0.30/m^3. The first block price is $0.27/m^3 and the highest is $0.40/m^3. As shown in Table 2.5, the average subsidy is about 57%. Observe that all domestic users benefit from subsidy. Fifty-three percent passing from 26.2 DA/m^3 (0, 37$) to 40 DA/m^3 (0, 40$). For domestic use, the increase is about 40% (22.2–31 DA/m^3) (see Table 2.3).This indicates that the water in Algeria is priced far below its cost (Drouiche et al. 2012).

2.7.10 Public–Private Sector Partnership and Water Utility Management

Suez Environment was awarded a water management contract for Algiers in November 2005 for 5.5 years. Its objective was to distribute water for several purposes and improve customer satisfaction. As a result of good services in 2009, the reliability of supply increased from 16% to 80%, seven beaches were opened to the public, and extensive training has been conducted (Suez Environment/SEAAL 2009). In 2011, reliability reached 99% and the contract was renewed for 5 more years.

Magtaa reverse osmosis desalination plant is constructed on a DBOOT (design-build-own-operate-transfer) basis. Under the Hyflux agreement, the project company supplies 500,000 m^3 a day of desalinated water to L'Algerienne Des Eaux (ADE), a state-owned national public water entity in Algeria. HWD or "Hamma Water Desalination Plant" is also a desalination plant on BOO (build-own-operate) basis. It is a joint venture between GE and Algerian Energy company (AEC) which supplies 200,000 m^3/day of water to Algeria (Drouiche et al. 2011).

2.8 Water and the Environment

Effective management of water resources demands a holistic approach, linking social and economic development with protection of natural ecosystems. Ecosystems of the earth are related and maintained by water. Water drives growth of plants and provides a permanent habitat for many species (Young et al. 2004).

2.8.1 Water and Ecosystems

Water plays a vital role in ecosystems across the earth. Although many other substances are necessary for life and for ecosystems to survive, without water nothing else would function to sustain life. The dry climate of Algeria is indicative of the exploitation and reduced recharge of groundwater. Ecosystem management is one of the main principles of water management for people and the environment as suggested by Acreman (1998). Fig. 2.4 demonstrates the implications of using water for humans in an indirect way, by supporting ecosystem processes and using it directly for humans. The upper part of Fig. 2.4 shows a positive impact of utilizing water for natural ecosystems that provide valuable goods, services, and tourist value. It suggests that every form of life is unique in its own way and all should be taken care of to maintain the essential ecological processes to receive benefit from the nature. The lower part of Fig. 2.4 shows the effect of direct use of water for the welfare of society, which has benefited society but at times is also a cause of pollution. Therefore, a tradeoff is to be maintained to decide direct water use by people for domestic use, agriculture, and industry and indirect water use by people to maintain ecosystems, which are responsible for providing environmental goods and elemental services.

2.8.2 Brine Disposal

Seawater desalination is growing rapidly, with many plants in operation, which affect coastal water quality. This is mainly due to the highly saline brine that is released into the sea, which may be increased in containing residual chemicals from pretreatment process, cleaning agents, and heavy metals from corrosion. The desalination concentrates from desalination plants vary widely, with multiple effects on water, biotic community, sediment, and marine organisms. It, therefore, threatens the quality of marine resources.

There are few actual regulations, standards, or guidelines for brine discharges around the world. In Algeria, brine disposal is dealt with in two different methods, in seawater plants, submerged disposal, by means of pipe that transports brine far into the sea to minimize concentration and facilitates faster and greater dilution, and evaporation ponds in case of brackish water. Both methods intend reducing its environmental impact (Mooij 2007; ICWE 1992). Diffusers are used in the pipelines to dilute the brine. The desalination plants at Beni Saf and Mostaganem in Algeria demonstrate the use of such diffusers (Mooij 2007; Algerian law No. 83–17; Decree No. 93–160).

2.9 Energy–Water Nexus

The water–energy nexus is the relationship between water used to generate and transmit energy and the energy required to collect, clean, move, store, and dispose water. A pictorial representation of energy–water nexus is shown in Figs. 2.5 and 2.6

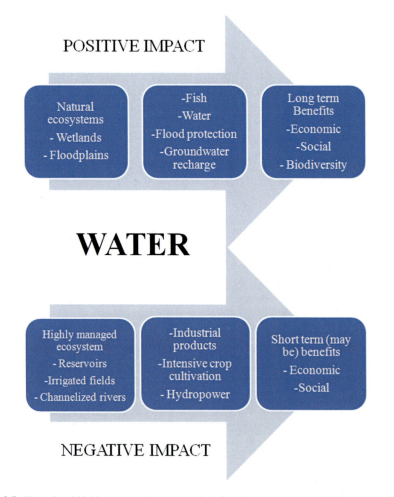

Fig. 2.5 Natural and highly managed ecosystem benefits. (Source: Acreman 1998)

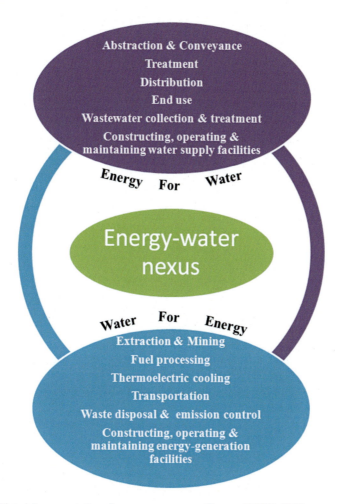

Fig. 2.6 Pictorial representation of energy water nexus. (Source: IRENA 2015)

Predictions regarding climate change indicate that rainfall might decrease by more than 20% by 2050, resulting in the worsening of water shortages in different parts of Algeria. The development of nonconventional water resources to meet the demand has increased the energy consumption of the water sector. In Algeria, the water sector in 2011 consumed around 4983 GWh and is set to raise to 16,090 GWh by 2030, more than treble (Hamiche et al. 2015). This predicted increase is attributed mainly to seawater desalination, water transfer projects, supply of water through pipes, and wastewater treatment facilities. As of 2014, Algeria's energy mix is mainly based on natural gas (more than 90%) in terms of power generation. Natural gas reserves are expected to last for 30 more years, that is, until 2030 (Hamiche et al. 2015). By 2030 it is expected that 30–40% of the electricity produced for domestic consumption will be from solar energy.

While there are many impediments and challenges toward NEXUS approach, these can be overcome by comprehensive planning, risk assessment, and policy implementation. Undoubtedly, the holistic Nexus approach in Algeria will empower a hub for knowledge and technology exchange and for innovating, adding value to economy by providing employment opportunities, reducing environmental impacts, and adapting and benchmarking solutions (Drouiche and Aoudj 2015). As energy production is based on fossil fuels, a finite source, it is obvious that promoting renewable energies to power desalination plants is needed. To further avoid water scarcity a coordination is required between the water and energy sectors. The use of renewable energy (most important is solar) must be promoted, which will help save fossil fuels, share water needs in a sustainable manner, and also be the reason of technical advancement.

2.10 Special Issues

2.10.1 International Water

2.10.1.1 Groundwater/Surface Water

Algeria has access to two main shared water basins: The North Western Sahara Aquifer System (NWSA), shared between Algeria, Libya, and Tunisia, and the hydrological basin of Bounaim-Taffna, shared between Morocco and Algeria.

The hydrological basin of Bounaïm-Tafna shared between Morocco and Algeria has two reservoirs: the Angad-Maghnia unconfined aquifer and the Jbel-Hamra confined aquifer. Due to overexploitation of groundwater, both countries suffer from quality and quantity of water resources. The unconfined aquifer has nitrate contamination (Boughriba et al. 2010), and in some places people use the untreated groundwater for drinking, which causes adverse health impacts. The water situation is stressed due to scanty rainfall and anthropogenic interferences.

The Medjerda river basin covers 23,700 km^2 and is shared between Tunisia and Algeria. The Medjerda River, which encompasses 22 reservoirs, represents 37% of the surface water of Tunisia and receives about 15,069 m^3/day of effluents. The sources of pollution are mainly used waters (95%); agriculture (0.14%); chemical industries such as plastic, automotive, textile, oil, and paper (0.44%), and food processing industries. Water quality is threatened by point source pollution, including municipal sewage discharges, industrial wastewater loads, and nonpoint source pollution from agriculture (Etteieb et al. 2017). An agreement was established between the Algerian and Tunisian governments to assure the management and access to the water resource and a joint technical commission for the Water Resources Planning and Management was formed for the exchange of information and data.

2.10.1.2 Conflicts, Negotiations, and Agreements

The South of Algeria, dominated by the Saharan platform, has large aquifers that cover hundreds of thousands of km². Algeria shares with Tunisia and Libya these immense reserves, which are only partially usable and weakly renewable. Due to the increased abstraction that caused a decline of the aquifer artesian pressure and groundwater salinization, an aquifer management organization was established to coordinate the plans and management. The three countries reached an agreement in 2002 to establish a Consultation Mechanism for the North-West Sahara Aquifer System (NWSAS). Under the umbrella of the Observatory of the Sahara and the Sahel (OSS), the three countries have initiated joint studies to assess the risks facing the Saharan Basin and to establish a consultation mechanism that is responsible for supporting the countries in implementing the main technical activities aimed at facilitating consultation, especially data collection, identifying transboundary water resources challenges, ensuring information dissemination, and organizing discussion at the level of decision makers for the NWSAS. The Consultation Mechanism was implemented in the form of a steering committee and was elaborated into a permanent structure in 2006 (Edmunds et al. 2003; Intergovernmental Panel on Climate Change (IPCC) 2007; Algeria et al. 2014).

2.11 Social, Equity, Institutional Performance, and Other Issues

Algeria has made good progress in extending water supply and sanitation coverage during the past two decades, under clear legislation and policies to three complementary sets of targets: the Economic Development, Millennium Development Goals (2015), and Strategy 2020. The institutional framework has been reinforced by the recently updated National Policy and the Water Law (2005), addressing all four subsectors. The Ministry of Water Resources leads coordination of stakeholders in the water supply subsectors, sharing this role with the Ministry of Health in the case of sanitation. However, there are outstanding challenges regarding planning and budgeting, monitoring and evaluation, as well as capacity building at the lower levels of the government following decentralization.

According to statistics supplied by ADE, indicators on network connection reached 98% nationally, the consumption exceeds 200 L/day/capita in 27 wilayas (state), with a national average of 180 L/day/capita. The same utility indicates that 73% of Algerians receive water daily. However, in addition to the disparity between regions, there is the problem of the intermittent water distribution in 62% of the Algerian households. This phenomenon, which affects households as well as administrations and companies, results in additional expenditure on water storage equipment and electricity bills that are not always easy to bear.

However, since water is poorly distributed, dissatisfaction is often expressed in several regions. According to the Ministry of Water Resources, Algeria has storage capacity of 8 billion m^3 from the existing 75 dams, to which nine new dams will be added, by 2018, with additional capacity of 500 million m^3. Also, the desalination stations provide 2.1 million m^3/day (Maghreb Emergent 2011).

2.12 Conclusions

Algeria relies on scarce water resources unevenly distributed. Due to scanty rainfall and drought persistent situations, the level of reservoirs and dams is fluctuating, thereby causing a negative impact on social and economic activities in the country. On the contrary, population growth and development in the country have increased the demand of water for drinking, industrial, and agricultural purposes. In recent years, Algeria invested heavily to address water scarcity. The main actions performed consisted of reducing water losses, building new dams, rationalization of aquifer use, popularizing the use of treated wastewater, and desalination of seawater. Algeria is faced by several challenges in the water sector such as the climate change and lower rain fall, leakages in the old water supply system, water pollution due to increased urbanization, unauthorized drilling of wells and boreholes, and poor water quality in the Sahara region.

Owing to the driving demand of water in Algeria, the country will have to depend upon nonconventional water since renewable water resources are overexploited. Other options to bridge the water gap are the rationalization of virtual water trade and use of renewable energy for desalination. The quality of rivers is deteriorated by various forms of pollution. Water resources have become increasingly limited, difficult to exploit, and often are exposed to significant amounts of wastewater.

Government efforts alone cannot improve the efficiency of Algeria's water supply. Private investments are necessary to address the country's water challenges. Public–private partnerships in the water sector are partly responsible for the recent surge in desalination capacity. In addition, private organizations are responsible for water management in some of the country's largest cities.

The water policy reforms show progress and address the challenges of water scarcity through an investment strategy, institutional strengthening, and legislative and regulatory reforms. But due to the increasing water demand for various purposes, there is still a need for efficient use, management of water resources, and public awareness of the importance of water conservation. It is imperative for the authorities to implement the principles of sustainable and integrated resource management.

Researchers contribute to Algeria's water knowledge base by tackling fundamental water research, growing scientific capacity, and disseminating knowledge to important stakeholders. Research-funded projects directly address the country's water challenges by investigating new technologies and methods to enhance water and sanitation supply, supporting policy and legislation, and providing much-needed guidance to decisionmakers. The research topics cover all aspects of the water cycle, including water resource management, aquatic ecosystems, desalination, water use, waste management, and the use of water in agriculture. They also consider climate change that will affect the water resources in the future.

Several agreements to strengthen bilateral cooperation in water field with countries facing the same challenges (i.e., droughts, flooding, and other events exacerbated by climate change) were signed aiming at facilitating the inflow of relevant results from research projects to the water sector and promote collaboration. Moreover, The European Union and Algeria are strengthening their scientific collaboration under several research and innovation programs in order to better tackle issues related to water research and innovation.

References

Acreman, M. (1998). Principles of water management for people and the environment. In A. de Shirbinin & V. Dompka (Eds.), *Water and population dynamics* (321 pp). Washington, DC: American Association for the Advancement of Science.

Agence du Bassin Hydrographique Constantinois-Seybousse-Mellegue. (2009). *Institutional framework and decision-making practices for water management in Algeria.* Towards the development of a strategy for water pollution prevention and control in the Seybousse River Basin.

Algeria MWR, CEDARE, Denmark, A. (2014). Algeria 2012 state of the water report, Monitoring & Evaluation for Water in North Africa (MEWINA) project, Ministry of Water Resources, Algeria – MWR, Water Resources Management Program – CEDARE.

Algerian Energy Company (AEC). (2011). Website: www.aec.dz

Algerian law No. 83-17 — relating to the water code.

Benmehaia, M. A., & Brabez, F. (2016). The propensity to cooperate among peasant farmers in Algeria: An analysis from bivariate approach. *International Journal of Food and Agricultural Economics (IJFAEC), 4*, 79.

Benmehiaia, M., & Brabez, F. (2016). Determinants of on-farm diversification among rural households: Empirical evidence from Northern Algeria. *International Journal of Food and Agricultural Economics, 4*(2), 87 99.

Bessenasse, M., Kettab, A., & Moulla, A. S. (2010). Seawater desalination: Study of three coastal stations in Algiers region. *Desalination, 250*, 423–427.

Boughriba, M., Barkaoui, A.-E., Zarhloule, Y., Lahmer, Z., El Houadi, B., & Verdoya, M. (2010). Groundwater vulnerability and risk mapping of the Angad transboundary aquifer using DRASTIC index method in GIS environment. *Arabian Journal of Geosciences, 3*, 207–220.

Boye, H. (2008). *Water, energy, desalination and climate change in the Mediterranean.* Sophia Antipolis: Plan Bleu.

CEDARE. (2014). Algeria Water Sector M&E Rapid Assessment Report. Monitoring & Evaluation for Water in North Africa (MEWINA) project, Water Resources Management Program, CEDARE.

Cheurfi, W., Bougherara, H., Bentabet, O., Batouche, K., & Kebabi, B. (2009). Fighting against nitrate pollution of the dam-retained waters through biological treatment. *Scientific Study & Research-Chemistry and Chemical Engineering, Biotechnology, Food Industry*, 285–294.

da Fonseca, A. F., Melfi, A. J., & Montes, C. R. (2005). Maize growth and changes in soil fertility after irrigation with treated sewage effluent. I. Plant dry matter yield and soil nitrogen and phosphorus availability. *Communications in Soil Science and Plant Analysis, 36*, 1965–1981.

Decree No. 93-160 — Regulating discharges of industrial liquid effluents.

Dinar, A., & Mody, J. (2004). Irrigation water management policies: Allocation and pricing principles and implementation experience. In *Natural resources forum*, vol 2. Wiley Online Library, pp. 112–122.

Drought management strategy in Algeria. (2014). Regional workshop for the Near East and North Africa.

Drouiche, N., & Aoudj, S. (2015). Water-energy-food nexus approach: Motivations, challenges and opportunities in Algeria. *International Journal of Thermal and Environmental Engineering, 10*, 11–15.

Drouiche, N., Ghaffour, N., Naceur, M. W., Mahmoudi, H., & Ouslimane, T. (2011). Reasons for the fast growing seawater desalination capacity in Algeria. *Water Resources Management, 25*, 2743–2754.

Drouiche, N., Ghaffour, N., Naceur, M. W., Lounici, H., & Drouiche, M. (2012). Towards sustainable water management in Algeria. *Desalination and Water Treatment, 50*, 272–284.

Easter, K. W., & Liu, Y. (2005). *Cost recovery and water pricing for irrigation and drainage projects agriculture and rural development* (Discussion paper 26).

Edmunds, W., Guendouz, A., Mamou, A., Moulla, A., Shand, P., & Zouari, K. (2003). Groundwater evolution in the Continental Intercalaire aquifer of southern Algeria and Tunisia: Trace element and isotopic indicators. *Applied Geochemistry, 18*, 805–822.

Etteieb, S., Cherif, S., & Tarhouni, J. (2017). Hydrochemical assessment of water quality for irrigation: A case study of the Medjerda River in Tunisia. *Applied Water Science, 7*, 469–480.

Export.gov. (2019). *Algeria – public works, infrastructure development, and water resources | export.gov.* [online] Available at: https://www.export.gov/article?id=Algeria-Public-Works-Infrastructure-Development-and-Water-Resources. Accessed 6 Mar 2019.

Falkenmark, M. (1986). Fresh water: Time for a modified approach. *Ambio, 15*, 192–200.

fao.org. (2015). *FAO AQUASTAT.* [online] Available at: http://www.fao.org/nr/water/aquastat/countries_regions/DZA/DZA-map_detailed.pdf

Foster, S., & Loucks, D. (2006). *Non-renewable groundwater resources*. Paris: UNESCO.

Ghaffour, N. (2009). The challenge of capacity-building strategies and perspectives for desalination for sustainable water use in MENA. *Desalination and Water Treatment, 5*, 48–53.

Ghazouani, W., Molle, F., & Rap, E. (2012). Water users associations in the NEN region: IFAD interventions and overall dynamics International Fund for Agricultural Development and International Water Management Institute, [Draft].

Ghodbane, M., Boudoukha, A., & Benaabidate, L. (2016). Hydrochemical and statistical characterization of groundwater in the Chemora area, Northeastern Algeria. *Desalination and Water Treatment, 57*, 14858–14868.

Good Practices in Agricultural Water Management Case Studies from Farmers Worldwide. (2005, April 11–22). *Commission on sustainable development thirteen session*. New York.

Groundwater Management in Algeria. Draft Synthesis Report. (2009). Food and Agriculture Organization of the United Nations Rome.

Groundwater Management in the Near East Region Synthesis Report. (2011). Food and Agriculture Organization of the United Nations Rome.

Hamdy, A., & Lacirignola, C. (2005). *Coping with water scarcity in the Mediterranean: What, why and how*. CIHEAM Mediterranean Agronomic Institute Bari Italy, pp. 277–326.

Hamiche, A. M., Stambouli, A. B., & Flazi, S. (2015). A review on the water and energy sectors in Algeria: Current forecasts, scenario and sustainability issues. *Renewable and Sustainable Energy Reviews, 41*, 261–276.

Hammouche, H. (2011). *Algeria report country experts consultation on wastewater management in the Arab world*. Dubai: The Arab Water Council.

Hirche, A., Salamani, M., Abdellaoui, A., Benhouhou, S., & Valderrama, J. M. (2011). Landscape changes of desertification in arid areas: The case of south-West Algeria. *Environmental Monitoring and Assessment, 179*, 403–420.

ICWE. (1992). *The Dublin statement and record of the conference. International conference on water and the environment*. Geneva: World Meteorological Organization.

Intergovernmental Panel on Climate Change (IPCC). (2007). *IPCC fourth assessment report: Climate change 2007 (AR4)*. Available from: http://www.ipcc.ch/publications_and_data/publications_and_data_reports.shtml

Investment Climate Statement – Algeria. (2015). *Bureau of economic and business affairs*. [online] Available at: https://www.state.gov/e/eb/rls/othr/ics/2015/241455.htm

IRENA. (2015). *Renewable energy in the water, energy and food nexus*. [online] Available at: https://www.irena.org/documentdownloads/publications/irena_water_energy_food_nexus_2015.pdf

Karrou, M., Oweis, T., & Bahri, A. (2011). *Improving water and land productivities in rainfed systems. Community-based optimization of the management of scarce water resources in agriculture in CWANA*. Report.

Kettab, A. (2001). Les ressources en eau en Algérie: stratégies, enjeux et vision. *Desalination, 136*, 25–33.

Knoema. (2014). *Algérie renewable surface water*. [online] Available at: https://knoema.fr/atlas/Alg%C3%A9rie/topics/Eau/Ressources-totaux-deau-renouvelables/Renewable-surface-water. Accessed 6 Mar 2019.

Laoubi, K., & Yamao, M. (2008). Algerian irrigation in transition; effects on irrigation profitability in irrigation schemes: The case of the East Mitidja scheme World Academy of Science. *Engineering and Technology, 48*, 293–297.

Maghreb Emergent. (2011). *Accueil > Maghreb Emergent*. [online] Available at: http://www.maghrebemergent.com/actualite/maghrebine/76492-62-des-algeriens-n-ont-pas-d-eau-24-heures-sur-24-ade.html. Accessed 6 Mar 2019.

Maliva, R., & Missimer, T. (2012). *Arid lands water evaluation and management*. Berlin: Springer.

Meddi, H., & Meddi, M. (2007). Spatial and temporal variability of rainfall in north west of Algeria Geographia. *Tech, 2*, 49–55.

Millennium Development Goals. (2015). *The Millennium Development Goals Report 2015*. United Nations 2015, ISBN 978-92-1-101320-7.

Ministère des Ressources en Eau. (2011). [online] Available at: http://www.mre.gov.dz/. Accessed 6 Mar 2018.

Mohammad, M. J., & Mazahreh, N. (2003). Changes in soil fertility parameters in response to irrigation of forage crops with secondary treated wastewater. *Communications in Soil Science and Plant Analysis, 34*, 1281–1294.

Mooij, C. (2007). Hamma water desalination plant: Planning and funding. *Desalination, 203*, 107–118.

National Policy and the water Law. (2005). *Loi relative à l'eau N° 2005-12 du 28 Joumada Ethania 1426 correspondant au 4 août 2005*. p. 3.

Office Nationale de l'Assainissement (ONA). (2011). [online] Available at http://www.ona-dz.org/

Remini, B., & Achour, B. (2013). The foggaras of in Salah (Algeria): The forgotten heritage. *LARHYSS Journal*. ISSN 1112-3680.

Righttowater.info. (2019). *Righttowater.info*. [online] Available at: http://www.righttowater.info/progress-so-far/national-legislation-on-the-right-to-water/#AL

Sekkoum, K., Talhi, M. F., Cheriti, A., Bourmita, Y., Belboukhari, N., Boulenouar, N., & Taleb, S. (2012). Water in Algerian Sahara: Environmental and health impact. In *Advancing desalination*. London: IntechOpen.

Suez Environnement/SEAAL. (2009). *An innovative public-private partnership in the environment sector: The Management Contract Of Societe Des Eaux Et De L'assainissement D'Alger (PDF)*. World Bank Water Week. Retrieved September 28, 2011.

Terra, M. (2011). *Algerian water policy: The potable water problem*. The first international seminar on water, energy and environment, ISWEE 11 Algiers, Algeria.

The world bank Report No. : 36270 – DZ. (2007). *People's democratic Republic of Algeria a public expenditure review.* [online] Social and Economic Development Group Middle East and North Africa Region. Available at: http://siteresources.worldbank.org/INTALGERIA/Resources/ALGERIAPER_ENG_Volume_I.pdf

Water Desalination Report. (2011). www.waterdesalreport.com

Young, G. J., Dooge, J. C., & Rodda, J. C. (2004). *Global water resource issues.* Cambridge: Cambridge University Press.

Nadjib Drouiche is Senior Researcher at the Centre de Recherche en Semi-Conducteurs pour l'Energétique (Algeria). He is the Director of the Crystal Growth and Metallurgical Processes (CCPM) and Head of the environmental team. His research interests include adsorption, membrane processes, electrochemical processes using sacrificial anodes, advanced oxidation processes, and recovery of by-products from industrial waste. He has published more than 80 papers in ISI-ranked journals with more than 1000 citations, and his h-index is 20. He was awarded TWAS-ARO YAS Prize 2012: "Sustainable Management of Water Resources in the Arab Region."

Rafika Khacheba obtained her master's degree in water and environmental management from the University of Limoges, France, in 2013, and an MSc in agricultural hydraulics from the National Agronomic School, in Algeria, in 2011. She then worked as Head of the wastewater reuse department at the National Sanitation Office of the Ministry of Water Resources in Algeria. Currently she is a PhD student working on the water economy in order to meet the objectives of sustainable development.

Richa Soni earned her PhD from Malaviya National Institute of Technology, India, in 2015. Then she worked as a National Post-Doctoral Fellow (NPDF-SERB) in the School of Engineering at Indian Institute of Technology, Mandi, India. The research during the Post-Doctoral work included synthesis of graphene-based composites for water treatment applications. She has a passion for research in the areas of water and wastewater treatment. Her research interests include surveying areas related to water problems, sample collection and analysis, lab-scale fabrication of water treatment units, adsorbent synthesis and mechanism, electrocoagulation, and water management.

Chapter 3
Existing and Recommended Water Policies in Egypt

Khaled M. AbuZeid

Abstract This chapter presents the 2017 state of water in Egypt. It provides information on the corresponding balance of water resources and uses in Egypt in light of the water scarcity situation. It provides views on the implementation of the water-energy-food nexus approach in Egypt. It also addresses the challenges facing Egypt's dependency on the transboundary water resource of the Nile, and the dependency of Egypt on the use of non-conventional water resources and reuse of agriculture drainage and wastewater. This chapter highlights the role of virtual water imports in achieving Egypt's food security. It touches upon the urban water tariff system. It reflects on existing water policies and provides recommendations for water policies that would achieve quick wins in the future.

Keywords Nile · Groundwater · Nubian Aquifer · Policy · Virtual Water · Reuse · Tariffs · Nexus

3.1 Introduction

Egypt's limited renewable water resources pose a restrictive challenge to its sustainable development. This challenge is exacerbated by the ever-increasing water demand and population growth. This requires additional water resources especially for the municipal, agriculture and industrial sectors. The increasing competition over the Nile River waters among upstream countries, in the absence of a commonly agreed river basin management vision, threatens the sustainability of the existing water uses of Egypt in the future.

K. M. AbuZeid (✉)
CEDARE, 2 ElHegaz Street, Heliopolis, Cairo, Egypt
e-mail: kabuzeid@cedare.int

© Springer Nature Switzerland AG 2020
S. Zekri (ed.), *Water Policies in MENA Countries*, Global Issues in Water Policy 23,
https://doi.org/10.1007/978-3-030-29274-4_3

Table 3.1 2017 Estimated water balance for Egypt

Water resources	BCM/year	Water uses by sector	BCM/year
Use from primary water resources			
Nile River	55.5	Domestic	10.75
Non-renewable groundwater	2.1	Industry	5.4
Rainfall	1.30	Agriculture	61.6
Saline water desalination	0.35	Evaporation	2.5
Total	58.24		
Reuse from secondary water resources			
Nile Valley & Delta groundwater	7.5		
Agricultural drainage reuse	9.31		
Treated wastewater reuse	4.19		
Total	21		
Total water availability	80.25	Total uses	80.25

Modified from Egypt's Vision 2030, revised draft on the Water Resources Sector

According to the Egyptian Census, the population of Egyptians living in Egypt reached about 95 million people in 2017 with an average growth rate of about 2.56% in the previous 10 years (CAPMAS 2017).

The following table shows different water resources contributing to different water uses for Egypt as of 2017 (modified from Egypt's Vision 2030 Sustainable Development Strategy) (Table 3.1).

Urban water demand is rapidly increasing and posing pressures on water resources due to population growth and urbanization expansion where the urban population has increased from about 41 million people in 2006 to about 55 million people in 2017 (CAPMAS 2017). Sanitation services coverage remains a challenge. Although about 97% of the households are connected to national potable water networks, only about 56% of the households are connected to national sewage networks (CAPMAS 2017).

The impacts of climate change on the water sector in Egypt need assessment not only at the national level but also at the transboundary level where 97% of the renewable water resources of Egypt originate upstream in the Nile River basin.

From the institutional perspective, the water sector in Egypt is currently divided among two main Ministries: the Ministry of Water Resources and Irrigation, in charge of water resources planning and management, irrigation, and agriculture drainage, and the Ministry of Housing, Utilities and Urban Communities, in charge of domestic water supply and sanitation. Egypt has several water legislative frameworks and laws governing the water sector, mainly the 1984 Law 12 for Irrigation and Drainage, and the 1982 Law 48 for the Protection of the Nile and Waterways from Pollution. Current modifications to these laws are underway, and a New Law for Domestic Water has also been drafted.

3.2 Agricultural Water

The agriculture sector in 2017 used about 61.6 billion cubic meters (BCM) of water. However, a substantial amount of about 21 BCM/year came from recycled water in the form of agriculture drainage, treated wastewater, and return groundwater recharge.

(a) Surface water

About 39.2 BCM/year are withdrawn for agriculture purposes from fresh surface waters of the Nile River and its irrigation network of about 33,550 km in length, in addition to the effective rainfall in the amount of about 1.3 BCM/year.

(b) Groundwater

About 2.1 BCM/year are withdrawn from non-renewable groundwater aquifers for agriculture purposes. In addition, about 5.5 BCM/year are abstracted from return groundwater recharge within the Nile Valley and Delta aquifers.

(c) Reuse of agriculture drainage and treated wastewater

About 9.31 BCM/year of agriculture drainage is being reused in agriculture, in addition to about 4.19 BCM/year of treated wastewater which is directly or indirectly reused in agriculture.

Most of the Nile waters in Egypt are used and recycled several times. As shown in Fig. 3.1, the two regulators/barrages, Edfina barrage on the Rosetta

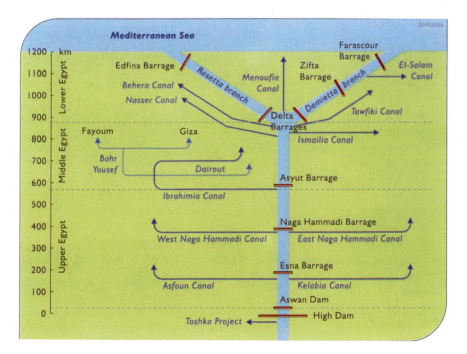

Fig. 3.1 Nile River dam/barrages system in Egypt. (Source: MWRI 2004)

branch of the Nile and the Faraskour barrage on the Damietta branch, provide full control on the tail ends of the Nile into the Mediterranean Sea whereby no freshwater is released from the Nile River except occasionally for environmental purposes. On the other hand, poor quality recycled agriculture drainage water, mixed in most cases with industrial and municipal wastewater, is discharged into the Mediterranean Sea.

Upper Egypt's agriculture drainage, in often cases still, finds its way to the Nile system and may be reused again downstream after mixing with fresh Nile water. Several agriculture drainage reuse pumping stations in the delta are located on agriculture drains to pump drainage water of good quality into irrigation canals for reuse to augment irrigation water, to meet tail end water demands. In cases where water qualities of these agriculture drains are affected by wastewater, and where their water quality is not meeting the set standards, reuse pumps are not operated. Unofficial reuse, where farmers at the tail ends of irrigation canals use their private mobile pumps on agriculture drains to irrigate their fields in the absence of adequate freshwater, also takes place (MWRI 2004).

(d) Cost Recovery

Irrigation water delivery costs are indirectly recovered through agriculture land taxes. However, this represents a partial cost recovery mechanism of the large-scale infrastructure used for irrigation water delivery. It has to be noted though that this infrastructure serves also for domestic, industrial, commercial, and other uses. On the other hand, irrigation costs at the farm level are mostly covered by the farmers or agriculture investors who cover their own costs of farm-level irrigation infrastructure, fuel for pumping, and other operation and maintenance costs. Costs of irrigation improvement mechanisms and agricultural drainage which are installed by the government are also recovered from the farmers in the form of installments.

(e) Institutional settings of irrigation organizations and farmers' participation

There are over 13,000 water users associations (WUAs) that are created so far in Egypt. However, these WUAs still do not cover the whole irrigation land. The new law, being discussed in parliament in 2018, aims among other things to set up WUAs as legal entities so that they can effectively take up their role in irrigation water management, including covering their own costs of canal and drainage maintenance at the farm level, and possibly farm irrigation improvement projects, and operation and maintenance costs of central pumps for irrigation rotation which serves more than one farmer.

(f) Irrigation efficiency

The Nile water system and its network of irrigation canals and agriculture drains in Egypt is considered one of the world's most efficient systems, with an overall water use efficiency, reaching over 75% in terms of water quantity. However, there is still room for improvement to increase the efficiency further at the local and farm level, and to improve the quality of water delivered, which is another aspect of water use efficiency. Irrigation improvement may add an additional 10–20% of freshwater availability. It will also provide for better distribution and more equity in allocating water quantity and quality among users, especially in the irrigation sector (AbuZeid 2011).

Reallocating any water savings that result from water efficiency programs has to be intelligently planned. Water savings from irrigation improvement projects, or domestic water supply networks rehabilitation, may not necessarily be reallocated to the same sector. With the ever-increasing demand in municipal water, a sector that has the highest priority in water allocation, it may be needed to reallocate Nile freshwater savings from the agriculture sector to the municipal water sector, and allocate the treated wastewater resulting from municipal water uses to agriculture expansion projects (AbuZeid 2009).

There is no detailed assessment of how much water has been saved from irrigation improvement projects, but so far, this has been contributing to the incremental increase in the supply of domestic water needed for urban expansion and the increasing population needs.

(g) State-of-the-art technology/Smart irrigation and innovations

A considerable area of desert reclamation projects for agriculture expansion is adopting modern irrigation schemes such as sprinkler and drip irrigation, and modern surface irrigation. Old agriculture lands in the Nile Valley and the Delta are being modernized. A little under 1 million acres have been modernized so far since the 1980s. Smart irrigation technologies are being implemented at experimental levels in few farms in Egypt.

(h) Food self-sufficiency vs food security (virtual water/food imports)

If Egypt would to be food self-sufficient, it would have to at least double its freshwater resources availability. That is why Egypt is adopting a food security strategy where it maintains a certain percentage (less than 100%) of food self-sufficiency in strategic crops and food products such as wheat, maize, table oil products, sugar, and rice. Egypt imports its remaining food needs and that is why it also has to maintain the availability of foreign currency needed to import its needs. Some of its high value cash crops exports contribute to the availability of this foreign currency. As per the Arab State of the Water Report 2015 (AbuZeid et al. 2019), Egypt was importing about 49 BCM/year of equivalent agricultural food products distributed as shown in Fig. 3.2, while exporting the equivalent of about 7 BCM/year in agricultural food products as shown in Figs. 3.2 and 3.3.

3.3 Domestic Water

Urban (Domestic) water is supplied at 10.75 BCM/year through surface water (about 8.4 BCM/year), groundwater recharge (about 2.0 BCM/year), and desalinated water (about 0.35 BCM/year in some coastal cities). Approximately 5.4 BCM/year is supplied to the industrial sector from surface water. Domestic water is the sector that has priority over other sectors in Egypt's water policies. In case of shortages, domestic water demand takes priority in satisfying demand over other sectors. The high dependency of Egypt on the one Nile River as the main source for renewable water resources requires long lengths of pipes to transfer domestic water to

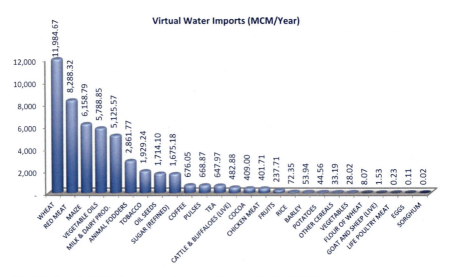

Fig. 3.2 Egypt's 2015 virtual water imports. (Source: AbuZeid 2017b)

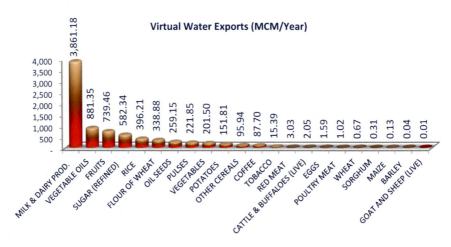

Fig. 3.3 Egypt's 2015 virtual water exports. (Source: AbuZeid 2017a, b)

urban centers of population and remote suburbs. The length of domestic water supply networks reached over 160,000 km as of 2015 (AbuZeid et al. 2019). These long transfers also require substantial amounts of energy for pumping. Sewage networks have also reached over 46,000 km in 2015 (AbuZeid et al. 2019).

(a) Desalination

All Red Sea tourism resorts in the South Sinai and the Red Sea governorates are already depending on sea water desalination for their water supply. All coastal cities are going to depend on desalination. Currently desalination provides about 0.35 BCM/year of domestic water. It is now the government strategy not to supply Nile water to coastal cities. Cities on the Mediterranean Sea and Red Sea coasts will depend mainly for their domestic water supply on sea water or brackish groundwater desalination.

(b) Wastewater treatment

As of 2015, annual produced municipal wastewater reached about 6.5 BCM. Industrial wastewater which is often collected using the same network has reached 4.2 BCM. 4.8 BCM of this total wastewater is being collected, and 4.25 BCM is treated annually.

(c) Wastewater reuse in the urban sector

As of 2015, 2.2 BCM/year of the treated wastewater is directly reused, and about 2 BCM/year is indirectly reused after disposal in agriculture drains.

(d) Urban water tariffs

The average domestic water tariff used to be about USD cents $7/m^3$ and it used to be about USD cents $2/m^3$ for the sanitation tariff. These tariffs did not meet the actual costs over the whole country, and some users were subsidized, either by cross-subsidy or by direct government subsidy. As of 2018, the Prime Minister's Decree No. 1012 for the year 2018 on the application of domestic water and sanitation tariffs for the fiscal year 2018/2019, household uses tariff for the first block from 1 to 10 m^3 was set at the equivalent of USD 3.75 cents/m^3. The second block from 11 to 20 m^3 was set at USD 9.22 cents/m^3 and the third block from 21 to 30 m^3 was set at USD 12.97 cents/m^3. The tariff for uses from 1 to 40 m^3 was set at USD 15.85 cents/m^3 and from 1 to more than 40 m^3 the tariff was set at USD 18.16 cents/m^3. Household sanitation tariff was set at 75% of the domestic water tariff.

Non-household uses were divided among several categories setting the tariff for "Services" at USD 19.02 cents/m^3, "Government" at USD 19.60 cents/m^3, "Commercial" at USD 20.75 cents/m^3, "Industrial" at USD 26.22 cents/m^3, "Tourism" at USD 26.51 cents/m^3, and "Others" at USD 34.58 cents/m^3, while "Sports and Social Clubs" at USD 57.64 cents/m^3. Sanitation tariff for these non-domestic uses was set at 98% of their water tariffs.

Drinking water tariff for border governorates such as "Sinai/Red Sea/ Matrouh" was set at a flat rate of USD 74.93 cents/m^3 and the sanitation rate was set at 50% of the domestic water tariff. The higher rate in these coastal governorates reflects their anticipated dependencies on desalination as the main source for urban water.

(e) Public–private sector partnership and water utility management

There are new laws in place to promote public–private partnership (PPP) and to regulate the business. The private sector is now involved in wastewater treatment projects that treat and provide treated wastewater for reuse in the cities. However, it is still at a small scale. It is still not that attractive to the irrigation sector due to low or no cost attributed to alternative irrigation water supplies. Also the weak law enforcement and protection of direct water-related investments such as new agriculture developments in desert lands are affecting Egypt's competitiveness in that area (AbuZeid 2011).

3.4 Water-Energy-Food Nexus

Innovations in dealing with the water-energy-food nexus need to be encouraged, especially when it comes to desalination, wastewater treatment technologies, and groundwater pumping. Currently, solar energy is being used for pumping groundwater in remote areas in the desert for irrigation purposes. Although this is contributing to food production and saving on fossil energy consumption and carbon emissions, it may be detrimental to the fossil groundwater reserves that is being pumped, if strong regulations and control on pumping are not enforced. This has to be studied using a nexus approach. On the other hand, some irrigation improvement policies, which may be envisaged as saving water, may actually be consuming unnecessary energy and not achieving the targeted water savings goals. The Ministry of Water Resources and Irrigation has alternatively been implementing a combined policy of improving surface irrigation in the old lands (within the Nile Valley and the Delta), while reuse agriculture drainage generated from improved surface irrigation, as well as enforcing modern irrigation such as drip and sprinkler irrigation in the new desert reclamation lands. This demonstrates that, when approached from a nexus perspective, transforming surface irrigation to pressurized drip irrigation is not always the best solution. It is important to look at the overall water use efficiency than just focusing on the on-farm irrigation efficiency (AbuZeid 2017a, b).

3.5 Special Issues

(a) *International waters*

(i) *The Nile River Basin*

Egypt is one of the 11 riparian countries sharing the Nile River basin, with South Sudan becoming the 11th Nile River basin country. Egypt is the most downstream country on the Nile River, which makes it vulnerable to upstream activities, even with the existence of bilateral agreements that protect its historical rights to a certain amount of the Nile waters. Reduced water availability could potentially affect Egypt's competitiveness in regional and world markets, especially in agriculture (AbuZeid 2011).

The Nile River basin receives 1660 BCM of rainfall per year. Egypt's share from the Nile River basin's water is 55.5 BCM/year. This share is documented by the bilateral agreement between Sudan and Egypt to share the average annual river discharge of 84 BCM/year that used to flow at Aswan, Egypt, before the construction of the High Aswan Dam. Sudan's share of that amount is 18.5 BCM/year, and the remaining 10 BCM/year is an estimate of the evaporation and seepage losses from Lake Nasser upstream of the High Aswan Dam. The rest of the 1660 BCM/year of rainfall over the Nile basin is either used up within the Nile River basin by rain-fed agriculture, grazing land, natural vegetation, and forests, or it contributes to groundwater recharge, or it's lost to evaporation from ground and surface water bodies (AbuZeid 2009).

Whereas Egypt depends mainly on the Nile River waters, other Nile basin countries depend mainly on direct rainfall within and outside the Nile River basin. The Nile basin countries, almost all upstream of Egypt, receive about 7000 BCM/year of rainfall within their boundaries. One Nile basin country, Congo, has another major river that runs through its territory, namely, the Congo River. The Congo River discharges into the Atlantic Ocean about 1000 BCM/year. This is almost 20 times the annual abstractions from the Nile River in Egypt. There is no freshwater discharge from the Nile River into the Mediterranean, due to the full utilization of the released Nile flows from the High Aswan Dam. The Nile basin country of Ethiopia has over 12 river basins, other than the Nile River basin within its territories (AbuZeid 2010). Unlike upstream Nile basin countries, Egypt depends almost completely on the Nile River waters (Blue Water) as compared to upstream higher dependency on the Nile basin's direct rainfall waters (Green Water) (AbuZeid 2012). The overall efficiency of water use in Egypt is considered to be among the highest in the world mainly due to water recycling and reuse.

Significant volumes of water losses exist in the upstream countries of the Nile due to the large areas of swamps, especially in the Sudd area in Southern Sudan, and in the Baro-Akobo area in Ethiopia. Water saving projects that could be implemented in these areas to increase the yield of the River Nile are estimated at 18 BCM/year in the Sudd in Southern Sudan, and 12 BCM/year in the Baro-Akobo in Ethiopia. The Jongli Canal Project was estimated to increase the White Nile River flows in Sudan by 4 BCM/year in its first phase to be shared by Egypt and Sudan. Some 70% of this canal has already been jointly constructed by Egypt and Sudan until it was halted due to political unrest in Southern Sudan (AbuZeid 2009).

(ii) *The Nubian Sandstone Aquifer*

The Nubian Sandstone Aquifer System (NSAS) is a transboundary groundwater basin in the North Eastern Sahara of Africa. The international waters of this regional aquifer are non-renewable and shared between Chad, Egypt, Libya, and Sudan. The area occupied by the Aquifer System is 2.2 million square km; 828,000 km^2 in Egypt, 760,000 km^2 in Libya, 376,000 km^2 in Sudan, and 235,000 km^2 in Northern Chad (AbuZeid and ElRawady 2010).

Policies related to non-renewable groundwater have to take into consideration the cumulative withdrawals compared to the groundwater reserves, and strategies for planned drawdown within a certain time frame have to be put in place. Figure 3.4 shows Egypt's cumulative withdrawals from the Nubian Sandstone Aquifer System, since its groundwater development started from its Nubian and Post-Nubian subsystems (PNSAS) (AbuZeid 2018 and CEDARE 2014).

The four countries sharing the Nubian Sandstone Aquifer have agreed to a regional strategy not to exceed 1 meter of drawdown per year (CEDARE 2002).

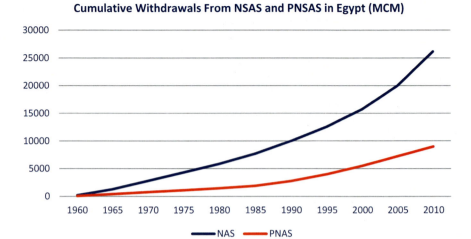

Fig. 3.4 Egypt's cumulative withdrawals from the Nubian Aquifer. (Source: CEDARE 2014)

(b) *Conflicts, negotiations, and agreements*

Egypt has water-related agreements with 5 of the 11 countries sharing the Nile basin. Egypt has initiated the Nile Basin Initiative (NBI), which was an umbrella framework that used to include 10 countries of the basin, with Eritrea being an observer. The NBI had two tracks: a joint projects development track and a legal framework development track. However, because of the unilateral signature of six countries on the Nile Cooperative Framework Agreement (CFA), before final agreement of all riparian countries on all articles, Egypt and Sudan withdrew from the NBI in 2010. The tensions in the relations between Egypt and Ethiopia (source of 85% of the Nile Waters running downstream) were exacerbated further in 2011 due to the unilateral decision of Ethiopia to build the Grand Ethiopian Renaissance Dam (GERD) on the Blue Nile, the largest tributary of the Nile supplying about 90% of Egypt's share of the Nile, with a storage capacity reservoir of exactly the same as Egypt's and Sudan's shares combined (74 BCM).

Studies have shown potential impacts on Egypt's and Sudan's shares from the Nile due to the accumulative effects of evaporation and seepages losses from the GERD. These accumulative effects would be much felt during the dry years and low flows of the Blue Nile. Figure 3.5 shows one of the several scenarios simulated in a study conducted by AbuZeid, K. at CEDARE.

A declaration of principles was signed in 2015 by the Heads of States of Ethiopia, Sudan, and Egypt on the GERD, mainly to agree on the rules of the first filling and the operation rules of the GERD before the first filling, and to oversee the environmental, socioeconomic, and hydrological Impact Assessment that is being

3 Existing and Recommended Water Policies in Egypt

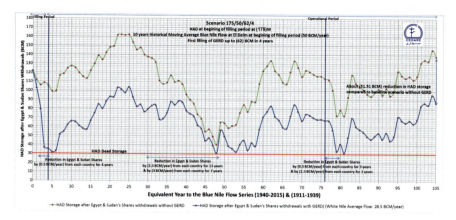

Fig. 3.5 Potential impacts of GERD on Egypt's and Sudan's Nile shares. (Source: AbuZeid 2017a, b)

conducted by an independent consultant that the three countries had agreed to hire. It is evident that this assessment should have been jointly conducted before the design and construction of the GERD, but the unilateral decision of Ethiopia prevented that from happening. It was announced in late 2018 by the Ethiopian government that due to delays, changes in the contractor, and in the design, it is supposed to be completed in 2022.

On the other hand, Egypt is part of the Joint Mechanism for the Studies of the Nubian Sandstone Aquifer with the three other riparian countries sharing the Nubian Sandstone Aquifer, namely, Libya, Sudan, and Chad.

(c) *Sustainability of non-renewable groundwater activities*

Non-renewable groundwater in most of Egypt's western desert, within the Nubian Sandstone Aquifer, which is also transboundary with Libya, Sudan, and Chad, is slowly being depleted. The aquifer is the only source of water and livelihood in the western deserts of Egypt covering the largest part of the country, almost 82%. The rate of depletion will be higher if this water is used for agriculture. The aquifer will last longer if this water is used for water bottling or used for municipal purposes. The quality of this groundwater is relatively high and requires minor treatment, therefore may be more appropriately used for human consumption. The return wastewater from municipal use of this non-renewable groundwater could be directed to agriculture after treatment, making this a far more efficient use of finite and non-renewable groundwater resources (AbuZeid and ElRawady 2010). According to the Holding Company for Water and Wastewater, 96.6% of the collected wastewater was safely treated in 2018. In 2017, about 350 MCM/year of treated wastewater was directly reused in agriculture, and about 3.5 BCM/year of treated wastewater was disposed into irrigation canals and agriculture drains where they were indirectly reused.

(d) *Property rights*

Groundwater users should have a license with specified allowed abstraction. The allowable pumping could be reduced based on the depletion situation of the groundwater aquifer. Egypt's lack of surface and renewable groundwater registered water rights reduces the ability to have a secure and stable future for agriculture land owners and water-dependent investments.

3.6 Conclusions

A quick win for Egypt is to implement a mix of policies that encourage conservation and use of non-conventional water resources, improve water allocation and water accounting, and jointly develop new Nile water resources. Egypt must plan now to use the appropriate type of water for the appropriate use at the appropriate location (AbuZeid 2009). Applying the fit-for-use water concept is going to result in reduced costs and efficient overall water use. Although there are some attempts to make use of treated wastewater for landscaping, and agriculture in some areas, and use of high value non-renewable groundwater for water bottling, which contribute to the concept, there is still no policy that is adopting the concept at the national level.

(a) Non-conventional water resources and geographical reallocation

Coastal cities on the Mediterranean and the Red Sea should be the first to consider desalination and start planning for it, if new water resources are to be sought. In the near future when Egypt will need to expand in desalination, proximity of seawater and brackish groundwater will play a role in reducing the cost of providing desalinated water. As it will be prohibitively expensive to convey desalinated water inland for long distances, inland governorates will therefore have to depend on Nile water, groundwater, and recycled water as the main sources for sustainable development in the future. This may require Nile water reallocation from some of the coastal governorates to inland governorates, which will be difficult to do in the future, if more Nile waters continue to be directed to coastal governorates away from the Nile River. With the existing limited share of Egypt from the Nile waters, no more internal Nile waters should be reallocated to coastal cities, as desalination would be the most appropriate resource for these appropriate locations (AbuZeid 2009). Desalination may appear to be the easy solution for providing new water resources; however, in some cases there are other priority options that are more cost effective for making more water resources available, especially agricultural water pricing, water conservation, and reuse of treated wastewater options. The following figure shows different incremental costs of supplying water for the city of Alexandria (AbuZeid et al. 2011) (Fig. 3.6).

(b) Facing the challenges: sectoral reallocation of water

Current depletion rates of renewable and non-renewable groundwater need to be addressed. There are weak monitoring and enforcement systems and low levels of awareness that result in illegal abstractions and over pumping of groundwater.

3 Existing and Recommended Water Policies in Egypt

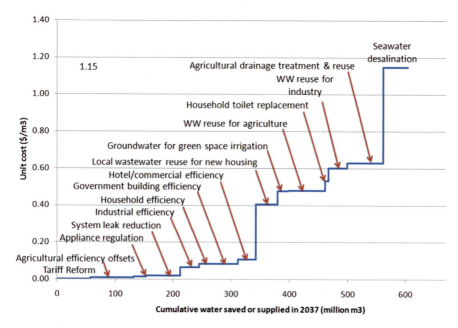

Fig. 3.6 Incremental cost for satisfying Alexandria future water demand. (Source: AbuZeid et al. 2011)

These may jeopardize the future of the existing investment in the agriculture sector and reduce significantly Egypt's competitiveness in the agriculture export marketing, and in achieving national food security. A reallocation of a different type of water resource may be needed to compensate for the depleted groundwater in the agriculture sector and to maintain the economic and social activities associated with the sector. However, reallocation of water away from the existing establishments in the agriculture sector is not recommended, due to the large socio-economic impact this might have on the country. A paradigm shift is needed to take the tough and wise decisions, and to get into innovative solutions at the technological, institutional, financial, and legislative levels, using non-conventional water resources such as treated wastewater, introducing models of public–private partnerships, and modifying the legislations that govern the way water is currently being managed to adapt to a more water-efficient and more competitive Egypt in 2030.

(c) Bridging the supply–demand gap

Water demand management, also known as water conservation, is a priority. There is room for reducing agriculture water losses by about 10% and domestic water losses by about 30%. A "Swap" in the type of water provided for agriculture represents another option of reallocation, whereby agriculture drainage water, or treated wastewater, rich in nutrients, may replace fresh Nile water for agriculture to free fresh Nile waters for drinking and domestic uses. This requires more attention to be given to the quality of agriculture drainage water, mainly by preventing the disposal of inappropriately treated or untreated wastewater into agriculture drains. It also requires raising the level of wastewater collection and treatment.

(d) Water quality improvement and reuse

Industrial wastewater treatment and recycling before disposal should be enforced to protect water quality. The modification of the wastewater reuse code that took place in 2015 to allow for reuse of treated wastewater in agriculture of food crops according to the level of treatment and the type of crop need to be embraced and fully implemented. This will contribute not only to food production but also to the improvement of water quality and the environment.

(e) Water security

Medium- to long-term measures include serious cooperation with the Nile basin countries to realize concrete win-win projects, such as those which provide more yield to the Nile waters, more hydropower, and more food for all, without affecting the existing uses in downstream countries such as Egypt.

(f) Institutional water reforms

A paradigm shift is needed in the way water is managed. Treated wastewater will need to be considered the renewable water resource for the future of agriculture (AbuZeid 2008). As wastewater increasingly becomes an important source, the domestic water and wastewater responsibilities (currently under the Ministry of Housing) may need to be brought together with the Water Resources planning responsibilities (currently under the Ministry of Water Resources and Irrigation). Every drop of water will need to be accounted for and measured, because "what is not measured cannot be managed" (AbuZeid 2009). Water should not be bought or sold as an economic good, but a cost-recovery model needs to be in place where the social, economic, and environmental value of water is considered. Water consumption and use needs to be measured. A mechanism for water equity needs to be established to avoid water conflicts and to ensure social cohesion. A model for equity needs to be established to define which sectors are to receive what portions of water, and how the rights of future generations (especially in regard to groundwater) are factored into such a model. One must also factor into the model ways of achieving equity among sectors and within sectors, and especially to ensure affordability for low-income sectors of the society through targeted subsidies or cross-subsidies. Partial or full "Volumetric Cost Recovery" (AbuZeid and ElRawady 2008) of water delivery services in all development sectors will need to be an acceptable practice in Egypt. Partial volumetric cost recovery which entails a subsidized payment for the service of providing water for municipal and even for irrigation use in volumetric terms will provide a sense of ownership and reduce over usage and wastage while recovering part of the costs. Water use tariffs to recover the costs of operation and maintenance as a function of the volume of water consumed are a socially sensitive and cost-effective tool for water efficiency, even if these tariffs are partially subsidized. However, this requires a strong metering and monitoring program for individual households and farmers, which is sometimes a challenge due to the vast number of apartment buildings and small holdings of farm land.

3 Existing and Recommended Water Policies in Egypt

(g) Quick wins

Quick win measures for Egypt as it moves into a highly competitive market and into a green growth economy should include flagship programs for water awareness and policy advocacy, on issues such as irrigation efficiency improvement, wastewater reuse and recycling, domestic water demand management, water legislative reform, and enforcement.

Modification of the building codes is needed to enforce the use of water saving devices in new developments and retrofit these in older ones. Proper water accounting is needed, where water meters and other appropriate measuring devices are installed, and where water consumption is recorded and transparently communicated to the officials and to the consumers in all sectors. An Integrated Water Resources Management (IWRM) law needs to be formulated to encompass all scattered water-related legislations that may be contradicting. Water users associations need to be legally recognized and given an official role and mandate. The private sector's role in the water sector needs to be clearly defined and legally accepted. Continuous capacity building programs and investment in human resources in the water sector are needed.

A water-efficient Egypt by 2030 can increase the quantity and quality of water available for its people and for future economic growth by a mix of policies including water demand management and conservation, advanced use of non-conventional water resources especially treated wastewater, and desalinated water (when appropriate), implementation of the fit-for-use water allocation policy among sectors and regions, enforcement of a strong water accounting and monitoring system, and joint development of new Nile water resources in cooperation with the Nile basin countries, especially South Sudan, North Sudan, and Ethiopia.

These quick win policies are a mix of short-term actions and long-term initiatives that can start as soon as possible and continue throughout the next decade to ensure that the "Gift of the Nile" remains the gift that keeps on giving.

References

AbuZeid, K., (2008, November 16–19). Integrated water resources management & water demand management opportunities in the Arab region. 1st Arab Water Forum, Arab Water Council, Riyadh, Saudi Arabia.

AbuZeid, K., & ElRawady, M. (2008). Water rights & equity in the Arab region. INWRDAM, Amman, Jordan.

AbuZeid, K., (2009, January). Water resources planning for Egypt in 2050, Workshop for water resources plans for Egypt in 2050. Ministry of Water Resources & Irrigation, Aswan, Egypt.

AbuZeid, K. (2010). Water footprints, green water, blue water, and virtual water, and their relevance to the legal and technical aspects of transboundary water management. Egyptian water partnership workshop, Egypt.

AbuZeid, K., & ElRawady, M. (2010). Sustainable development of non-renewable transboundary groundwater: Strategic planning for the Nubian Sandstone Aquifer. UNESCO ISARM conference, Paris, France.

AbuZeid, K. (2011). Egypt competitiveness report, water chapter. Center for Competitiveness, Egypt.

AbuZeid, K., et al. (2011). Alexandria 2030 Integrated Urban Water Management Plan (IUWM) Strategic plan, SWITCH project, CEDARE.

AbuZeid, K. (2012, July). The watercourse/blue water & river basin/green water approach to win-win solutions in transboundary river basin management. *The Official Journal of the Arab Water Council, 3*(1), 1–15. ISSN:1996–5699.

AbuZeid, K. (2017a). Potential impacts of grand Ethiopian Renaissance Dam on Nile water availability for Egypt and Sudan. 4th Arab Water Forum organized by the Arab Water Council, November 2017, Cairo, Egypt.

AbuZeid, K. (2017b). Chapter 7: The water, energy, and food security Nexus in the Arab region, research and development to bridge the knowledge gap. In K. Amer, et al. (Eds.), *Water security in a new world*. Springer. https://doi.org/10.1007/978-3-319-48408-2_7.

AbuZeid, K. (2018). Egypt State of the water indicators, Sate of the water indicators workshop, CEDARE, Egyptian water partnership. Arab Water Council, December 2018, Cairo, Egypt.

AbuZeid, K., Wagdy, A., CEDARE, & AWC. (2019). 3rd State of the Water Report for the Arab Region - 2015. Water Resources Management Program-CEDARE & Arab Water Council (AWC).

CAPMAS. (2017). Egypt Census Report 2017, Central Agency for Public Mobilization & Statistics (CAPMAS).

CEDARE. (2002). Regional strategy for the utilisation of the Nubian Sandstone Aquifer system. Programme for the development of a regional strategy for the utilisation of the Nubian Sandstone Aquifer system, water management program. Centre for Environment & Development for the Arab Region & Europe (CEDARE).

CEDARE. (2014). 2012 Nubian Sandstone Aquifer State of the Water Report, CEDARE, MEWINA, 2014.

Egypt's Vision 2030. (2017). Revised draft on the water resources sector.

Egyptian Prime Minister's Decree No. 1012. (2018).

MWRI. (2004). Egypt National Water Resources Plan till 2017, Ministry of Water Resources & Irrigation, Egypt.

Prof. Abu Zeid earned his BSc in Civil Engineering from Cairo University and his MSc and PhD in Civil Engineering and Water Resources Management from Colorado State University. He has 30 years of experience being CEDARE Regional Water Director; North Africa Water M&E Regional Coordinator; North African Ministers Council on Water Technical Secretariat Officer in Charge; Arab Water Council Governing Board Member; Egyptian Water Partnership Secretary General; Global Water Partnership East Africa (GWP-EA) Chairman; MCSD Vice Chair; Member of the Arab League's Council of Water Ministers' Advisory Committee, Arab Water Security Strategy Advisory Team, Mediterranean Water Strategy Experts' Group, Arab Shared Waters Convention Consultative Group, and High Level Technical Committee of the Transboundary Waters Sector of the Program for Infrastructure Development in Africa; and Team Leader of the Nile Basin Decision Support System Conceptual Design, the 1st and 2nd Arab State of the Water Reports, the Nile Basin 2012 State of the Water Report, the Nubian Sandstone Aquifer 2012 State of the Water Report, the Alexandria Governorate 2030 Integrated Urban Water Management (IUWM) Strategic Plan, and the 2030 Egypt Wastewater Reuse Strategic Vision. He developed Egypt's 2050 Water Resources Policy Options and participated in the development of the Integrated Model for Egypt Water Resources Management and the Nubian Sandstone Aquifer Regional Strategy.

Chapter 4
Iran's Water Policy

Farhad Yazdandoost

Abstract Iranians have traditionally been harnessing water resources ingeniously and efficiently throughout their rich history; however, over the last four decades, the rapid pace of development, demographic expansions, overurbanization, disintegrated large agricultural expansions, and perhaps most importantly climate change and variation have taken their toll on a barely stable water resources supply/demand chain. As such, the country is now facing a water deficit in general and moderate to severe scarcity in many of its water-related development and mainstream operations.

Keywords Iran · Water resources · Water policy

4.1 Introduction

Iran is located in the west of Asia and covers an area of about 1.648 million km^2. Iran is bordered on the north by Armenia, Azerbaijan, Caspian Sea, and Turkmenistan; on the east by Afghanistan and Pakistan; on the south by the Oman Sea, the Strait of Hormuz, and the Persian Gulf; and on the west by Iraq and Turkey (Fig. 4.1). Climatologically, Iran is situated in the arid and semiarid regions of the world. Of the total area, 13% has a cold and mountainous weather, 14% has a moderate climate, and the remaining 73% is covered by dry weather.

The population of the country is estimated to be 79.11 million for the year 2015 with a population growth rate of 1.2%. Figure 4.2 shows the trend of population growth over the last century to date. There are higher concentrations of people in the north and the west. Tehran is the capital and the largest city. Other large cities are Mashhad, Tabriz, Esfahan, and Shiraz. The central parts of the country that are covered by deserts and salinas are rather uninhabited.

F. Yazdandoost (✉)
Department of Civil Engineering, K N Toosi University of Technology, Tehran, Iran
e-mail: yazdandoost@kntu.ac.ir

© Springer Nature Switzerland AG 2020
S. Zekri (ed.), *Water Policies in MENA Countries*, Global Issues in Water Policy 23,
https://doi.org/10.1007/978-3-030-29274-4_4

Fig. 4.1 Geographical situation of Islamic Republic of Iran

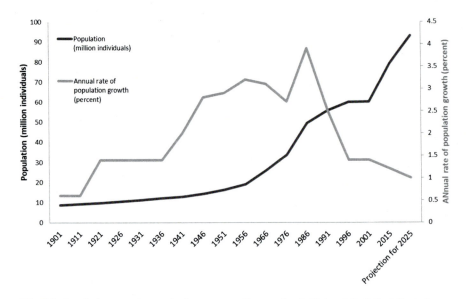

Fig. 4.2 Population growth over the last century. (Source: Iran's National Statistical Centre)

4.2 Water Trends

Iran is a mountainous country that is covered by mountains in more than 57% of its parts. Alborz and Zagross mountain ranges are the dominant features of the country with respect to water resources. Numerous plains and valleys spread in the spaces between the highlands through which several permanent and seasonal rivers flow.

There are some closed basins among main mountain ranges such as Lake Urmia (The largest natural, permanent lake in Iran), Arak Desert, Tashk and Bakhtegan Lakes, and Maharlu Lake.

Iran consists of six main hydrological catchments as follows:

- Caspian Sea Catchment, which covers northern part of Azerbaijan Province, northern slopes of Alborz, and some eastern and northern parts of Zagros slopes. All rivers in this region flow into Caspian Sea.
- Persian Gulf and Oman Sea Catchment covers Zagros in some main parts of its west and southwest heights and slopes.
- Lake Urmia Catchment, which covers the northern slopes of Zagros and eastern slopes of border mountains between Iran and Turkey as well as southern and western slopes of Mt. Sahand. All rivers in this area flow into Lake Urmia.
- Central Catchment, which covers all regions with flowing waters into central lakes, swamps, salinas, and deserts.
- North-eastern catchment named Gharaghoum Catchment and eastern catchment named Meshkil & Hamoun Catchment, which consist of those regions where water flow into border swamps and salinas and into Iran–Afghanistan and Iran–Pakistan borderlands.

The six main hydrological catchments of Iran are shown in Fig. 4.3.

Temporal and spatial distribution of precipitation in Iran is volatile, as 90% of total precipitation occurs in cold and humid seasons and in northern and western

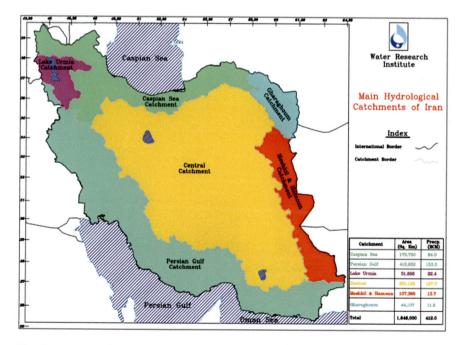

Fig. 4.3 Main hydrological catchments of Iran

parts of the country and only 10% occurs in warm and dry seasons and in central, southern, and eastern parts. About 52% of precipitation occurs in 25% of the area of the country; hence, several parts of the country will meet lack of water resources and water crisis in early future.

The average annual precipitation varies from 50 mm in central, southern, and eastern parts to 1500 mm in the western and northern parts of Iran. In other words, the average annual precipitation in 6% of the area of the country (eastern and central parts) is less than 50 mm, in 45% of the area (southern, eastern, and central parts) is less than 200 mm, in 40% of the area is 200–500 mm, in 8% of the area (northern and western parts) is 500–1000 mm, and in 1% of the area (north western coasts of Caspian Sea) is more than 1000 mm. Figure 4.4 shows the renewable water resources of the country.

Rainfall in Iran is highly seasonal, with a rainy season between October and March, leaving the land parched for the remainder of the year. Immense seasonal variations in flow characterize Iran's rivers. For example, the Karun River in Khuzestan carries water during periods of maximum flow that is ten times the amount borne in dry periods. In numerous localities, there may be no precipitation until sudden storms, accompanied by heavy rains, dumping almost the entire year's rainfall in a few days.

Water balance of the country according to 30 years of data shows that the average annual precipitation is 250 mm. Thirty percent of the precipitation occurs in the form of snow and the rest in the form of rain.

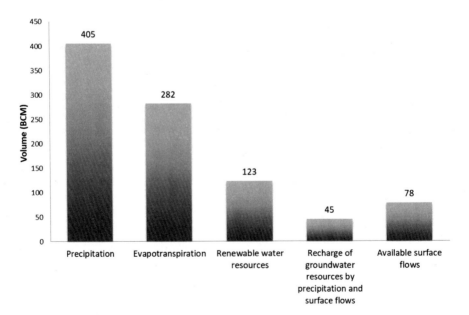

Fig. 4.4 Renewable water resources of Iran in 2016. (Source: Office of Water Resources Planning report, MoE)

Concerning the extent of the country, the average amount of annual precipitation is 405 billion cubic meters (BCM). A total of 282 billion cubic meters of it is lost as evapotranspiration, where 115 billion cubic meters is believed to be the share of direct evaporation. The surface flow stands at 78 billion cubic meters in the overall balance. Thirty-two billion cubic meters of infiltrated water recharges the groundwater resources and 13 billion cubic meters of water joins the groundwater from undersurface flows. So, the amount of renewable water is about 123 billion cubic meters per year, which is 30% of total precipitation (Yazdandoost 2016).

4.3 Agricultural Water

4.3.1 Surface Water

Iran is a country of contrasts with less than one-third of area suited for farmland, but because of poor soil and lack of adequate water distribution in many areas, most of it is uncultivated. Just 12% of the total land area is under cultivation and roughly one-third of the cultivated area is irrigated; the rest is devoted to dryland farming. Some 92% of agroproducts depend on irrigation. The western and northwestern portions of the country have the most fertile soils. Three percent of the total land area is used for grazing and small fodder production. Most of the grazing is done on mostly semidry rangeland in mountain areas and on areas surrounding the large deserts ("Dasht's") of Central Iran. The nonagricultural surface represents 53% of the total area with nearly 39% of the country not suited for agricultural purposes, 7% is covered by woodlands, and 7% is covered by urban areas (Karamidehkordi 2010). Table 4.1 shows consumption of water (with about 43% form groundwater) in the agriculture sector since 1963 up to 2015 in comparison with the population growth.

At the end of the twentieth century, agricultural activities accounted for about one-fifth of Iran's gross domestic product (GDP) and employed a comparable proportion of the workforce. Most farms are small, less than 10 hectares, and are not economically viable, which has contributed to the wide-scale migration to cities. In addition to water scarcity and areas of poor soil, seed is of low quality and farming techniques are antiquated. All these factors have contributed to low crop yields and poverty in rural areas (MoJA IR Iran 2014). In 2005, some 13 million hectares of land was under cultivation, of which 50.45% was allocated to irrigate farming. As of 2013, the amount of cultivated land that is irrigated increased to 8 million hectares, while a large portion remains rain-fed.

Table 4.1 Population and consumption changes in the agriculture sector during 1963–2015

Consumption	Year						Average annual growth (%)
	1963	1979	1988	1996	2001	2015	
Agriculture (billion cubic meters)	44.1	49.8	70.6	81.4	86	109.9	1.77
Population (million)	23.4	36.4	53.4	59.9	65	79.1	1.2

4.3.2 Groundwater

The internal renewable water resources are estimated at 123 billion cubic meters (BCM)/year of which some 33.5 BCM/year is the share of groundwater (MoE IR Iran 2016). It is estimated that groundwater resources provide more than 43% of agricultural demand. The groundwater abstraction is 63.8 BCM annually, which implies that nationally there is overexploitation of groundwater by 5.6 BCM, given the total infiltration of about 58 BCM, including returns from water consumptions. Groundwater abstraction (through wells, qanats, and springs) varied from less than 20 BCM annually in the early 1970s to more than 64 BCM in 2008 (Table 4.2). The number of wells during this period increased fivefold, from just over 9000 to almost 45,000 with Qanats number at about 34,000 and springs 56,000. Most of the overexploitation occurs in the central basins where less surface water is available. It is estimated that groundwater is the source of about 62% of all irrigated areas.

Although the national average long-term fall in the groundwater table has been about 0.5 m/year over the last decade, but more recent falls are greater than those presented in Table 4.2, indicating an even more severe condition. To avoid further groundwater depletion, some 210 of the 609 plains identified for further study are now heavily restricted in terms of withdrawal where well construction is not allowed in order to aid aquifers' recovery. The total area of aquifer under this management condition is 15,485 km^2 where the annual average depletion rate is 0.5 m and where the overabstraction totals some 4700 MCM/year. This estimate is greater than the depletion suggested in the FAO (2009) report and is provided by the Ministry of Energy (2016). Despite stringent controls, the agricultural sector is exerting increasing groundwater demands.

Table 4.2 Annual groundwater abstraction in the six major basins (MCM) in 2008

Basin	Wells	Qanats	Springs	Total	Average fluctuation in water table (m)
Khazar	4390	450	3000	7830	−1.58/−0.34
Persian Gulf and Gulf of Oman	10,080	1060	15,240	26,380	−1.69/−0.44
Lake Urmia	1970	230	120	2320	−0.96/−0.25
Markazi	25,930	5790	2470	34,180	−0.96/−0.57
Hamoon	800	400	60	1260	−0.25/−0.31
Sarakhs	1730	290	350	2360	−0.16/−0.00
Total	44,890	8230	21,240	74,350	

4.3.3 Legal Framework

4.3.3.1 Laws Governing Water Use in Agriculture

According to Iranian law, all water bodies are public property. The first water law after the revolution in the Islamic Republic of Iran was approved in 1982. Based on this law, allocating and issuing permits to use the water for domestic, agricultural, and industrial purposes are the responsibility of the Ministry of Energy (MoE). The Ministry of Agriculture (MoJA) is appointed to distribute water for agriculture among farmers and collect water fees. Numerous Water Users Associations have been established to assist with the public–private participation in this respect. Challenges with water resources management are exacerbated in this sector because of ownership issues, as traditionally land rights are confused with the embodied water rights.

4.3.3.2 Powers of Government Authorities in Charge of Administering Water in Agriculture

Based on Iranian law, water bodies (rivers, lakes, seas, etc.) are public property, and the government is responsible for their management. Article 1 of the Nationalization of Water Resources Act indicates that "All waters running in rivers, natural streams, valleys, brooks, or in any other natural courses, either surface or underground, as well as flood, sewage and drainage waters, and those of lakes, marshes, natural ponds, springs, mineral waters, and underground water resources are considered as national wealth and belong to the public, and the responsibility for the safeguarding and utilization of this national wealth and the establishment and management of water resources development establishments are charged to the Ministry of Water and Power."

Traditionally, the provision of water has been the responsibility of the government. As far as groundwater is concerned, the private sector invests in drilling wells, which are then operated and managed by farmers. In 1943, Iran set up an independent irrigation institute, whose function is to supervise and carry out all irrigation projects. It has the right to collect fees from the allocation of water and form societies through which private owners can participate in its work. In recent years there has been a large increase in private-sector financing of water projects, especially irrigation and drainage systems. Increased awareness of the limitations of natural resources has played an important role in adopting appropriate policies and strategies. Some of the current priorities are as follows: establishing a comprehensive management system over the whole-water cycle according to the principles of sustainable land and water development in river basins; developing water resources within the framework of national plans and comprehensive river basin plans; integrating water resources development, exploitation, and protection plans with other national and regional plans; promoting agricultural water productivity while remaining attentive to the economic, security, and political concerns related to the harvesting and extraction, supplying, storage, and consumption of water; ensuring that agricultural water resources are used not only efficiently but in a socially just manner; and promoting public awareness about limited natural resources.

4.3.3.3 Requirements for Licenses to Use Water

Based on Iranian law, the use of water resources requires obtaining a water use permit. No one is allowed to use water for any other purposes than what has been mentioned in the permit, nor is the permit transferable to others. A water use permit is applicable solely to the piece of land for which it has been issued, unless the government in the region decides otherwise and/or the use of the water is determined to be harmful or uneconomical. Since the early 2000s, renewed emphasis on integrated management of water resources has been witnessed, on the basis of which a new comprehensive water act and water management system are being developed (MoE 2016).

4.3.3.4 Providing Water for Agriculture and Settling Water Disputes

Groundwater is one of the most important water resources of Iran. One of the best methods of supplying water is digging *qanats* (subterranean canals), a practice that researchers consider was developed by Iranians about three thousand years ago. It must be noted that despite advancements in science and technology, there are still lessons to be learned from approaches deployed by our ancestors in harnessing water resources and observing natural harmony. This is clearly true with the case of management of limited water resources in arid–semiarid regions of the world where sustainable development was achieved in different historic periods in spite of challenging natural hardships.

In regard to surface water, any disputes over priority of use, quality, and quantity of water, as well as conflicts that cause delays in supplying, distributing, and consuming water should be settled in an arbitration committee by the water masters and chief water masters. However, if the disputes continue even after arbitration, the director of the region or the manager of the district, as the case may be, should intervene and investigate the matter and then give his recommendation.

Utilization of the groundwater resources through the drilling of any type of well or qanat anywhere in the country should be carried out with the permission of the Ministry of Energy, except in those cases where the drilling of such wells results in the depletion or drying up of the water in an adjacent well and/or qanat.

4.3.3.5 Ownership of Water Sources

Based on the Nationalization of Water Resources Act (1968), the development of water resources was to be supervised and controlled by the Ministry of Water and Power. Under the traditional irrigation systems, farmers receive their share of water based on their water rights, usually in proportion to the land area. For surface water, this right to water use is usually measured on the basis of the water delivery time. The water rights are attached to the land, and when the land is sold, the water rights are also transferred to the new owner. Eventually, based on the Civil Code of Iran, groundwater is generally regarded as private property and thus may be traded between farmers. Wells can be sold with or without the land. Qanats have shared ownerships. Those who have built a qanat or participate in its maintenance (rehabilitation and retrofitting) are entitled to partly use its water.

4.3.4 Reuse of Treated Wastewater

It is estimated that only 9% of wastewater is treated in Iran and treated wastewater provide about 2% of agricultural demand (MoE 2016). Secondary treatment is widely used for water demands by the municipalities.

4.3.5 Pricing

The average water productivity of agriculture (crops and orchards) is 0.7 kg per cubic meter (MoJA). Increasing the economic value of water is one of the major objectives in the Economic Development Programs of Iran. Based on the current law (passed in 1982), the price for regulated surface water is between 1% and 3% of the value of cultivated crops (Iran National Survey Organisation 2016). The water pumped from groundwater resources must be in accordance with the crop water requirement and proposed cropping pattern in each region. In this case, the price for groundwater is 0.25–1.00% of the commercial value of the crop yield. Lack of proper water pricing for agricultural purposes is one of the major obstacles for sustainable development.

4.3.6 Irrigation Efficiency

Modern agriculture has had a slow but steady growth in Iran, but a common perception is that irrigation water is wasted considerably in Iran. Studies have shown that the overall irrigation efficiency in Iran ranges from 33% to 37%, which is lower than average worldwide irrigation efficiency (MoJA 2014).

The Iranian Agricultural Engineering Research Institute (IAERI) conducted a research project on the on-farm water application efficiency for different crops (MoJA). The results obtained from field experiments in different provinces showed that the application efficiency depends on farm management, method of irrigation, growth stage, and type of crop, and it varied from 24.7% to 55.7%. Considering the conveyance efficiency in the target area of the research study, the overall irrigation efficiency varied between 15% and 36%. Considering overall irrigation efficiencies in different provinces and different published reports and research studies, the irrigation efficiency in the country is estimated to be 35%, which is very low compared to developing countries (45%) and developed countries (60%).

Half of the fully irrigated areas are equipped with modern irrigation systems and operated by governmental organizations (MoJA). Irrigation efficiency in such systems is very low and is measured at 20–30%. Irrigation efficiency may be so low due to subsidized water released from dams, there being no incentive for farmers to save water. The other half of fully irrigated areas are operated by the private sector, and the water is supplied from groundwater resources. In this case, the irrigation efficiency is also rather low and has been measured to be about 35%. The rest of the irrigated farms in Iran, which are under severe to moderate water stress, belong to small farm holders who do not save water, but their irrigation efficiency is quite high. Irrigation efficiency in these farms is estimated to be 55–65%. The reason for this may be due mainly to reduced irrigation, which they usually practice. These farmers have enough land, and the area under cultivation is much greater than the water available to them. They usually receive more benefit from their extensive farming with reduced irrigation in comparison with those who practice intensive farming and full irrigation.

4.4 Urban Water

4.4.1 Sources of Supply

Principal uses are categorized in different sectors in the country. Table 4.3 shows consumption changes in the urban water sector since 1963 up to 2015 in comparison with the population growth. It is estimated that surface water resources provide more than 60% of urban demand and groundwater resources provide more than 39% of it.

Similar to the agriculture sector, the urban water is also heavily subsidized by the government. It is estimated that the domestic water supply, with a national coverage of 99.4%, is charged at a much lower value than the actual cost of provision. Usage may stand at staggering figures of about 300 l per day per capita in some major cities and in particular in Tehran, the capital city. The government has recently focused on the problem of water scarcity by setting up a national task force for adaptation with water scarcity, whereby consumption/demand management is given the highest priority. Highest ranking officials from different departments form the national task force, while provincial authorities concentrate on allocation issues between the main users of water resources locally. Rearrangements and agreements on allocations between different sectors are reached on the basis of operational temporal and spatial needs and requirements.

Table 4.3 Population and consumption changes in the urban water sector during 1963–2015

Consumption	Year						Average annual growth (%)
	1963	1979	1988	1996	2001	2015	
Urban (BCM)	0.57	1.9	3.2	4.5	6	14.3	6.39
Population (million)	23.4	36.4	53.4	59.9	65	79.1	1.2
consumption per capita (m^3/y)	24.4	52.2	59.9	75	92.3	180.8	3.43

4.4.2 Drought Management

Planning and management of water in water distribution, quality, use, quotas, and rationing are practical methods to mitigate the effects of drought. An important issue regarding drought risk management is the "Agricultural Product Insurance Fund." In order to support vulnerable groups, especially farmers, extended loan reimbursement deadlines are considered by the banks. The mapping of areas at risk of drought, assessment of environmental resources (biodiversity) in areas coping with mid- and long-term drought (periods of 10–30 years), and provision of ecosystem sensitivity maps are drought preparedness strategies in emergency situations. Because of repeated droughts, the government implemented an aid and rescue program in 2003. The program included leadership and guidance, public training against natural disasters, duties of various institutions during crisis, and provision of mass media mapping; in addition, financial resources and statutory support during natural disasters were also considered. The National Water Programs focus on drought prevention and alleviation. Emergency interventions are introduced and aid measures include provision of agricultural inputs, small and medium funds, food subsidy, and forage.

4.4.3 Climate Change

The change of climate will cause a change in the hydrologic cycle. The change in the hydrologic cycle is likely to have a significant impact on water resources. There could be considerable changes in river flows, especially in arid and semiarid regions, thus increasing the risks in planning new projects (for instance, hydropower projects). Analysis of more than 30 years' data obtained from meteorological and hydrometrical stations in Iran indicates that out of 143 meteorological stations, 136 show undesirable climatic changes having a tendency towards arid climate, and out of 992 hydrometric stations, 36 show a decrease in normal discharge and an increase in the number of observed floods, which is the indicator of arid regions. During the recent years, the increase in Caspian Sea water-level elevation has inundated coastal areas. In the past 30 years (1985–2015), the temperature over Iran has increased about 0.6 °C on the average, and untimely snow melting and the decrease in precipitation have been observed generally. During the recent years, an increase in the frequency of destructive floods and droughts has also been recorded in hydrometeorological data. Research on the global warming effects on hydrology and water resources in Iran has been carried out in several rivers and lake basins by using historical hydrometeorological data and runoff models, together with global warming scenarios. The research has been carried out and comprehensive assessment work of global warming impact has been conducted to date. The main results obtained so far are:

- Based on different climate change scenarios, as the temperature rises, evapotranspiration will increase in most river basins throughout the year. A 2–6 °C increase in the temperature will augment the annual evapotranspiration by 6–12% in the six catchments throughout the country and decrease the annual runoff within the range of 1–5%.
- The modeled runoff distribution shows a significant increase in peak flood flow during the winter and a decrease in mean river discharge.
- Global warming will reduce the snowfall in winter with a significant change in seasonal pattern of river flow such as a decrease in the snowmelt-generated spring flow. This change affects various current water uses and the operation of existing facilities of water resources management.

4.4.4 Water Quality

Inadequate management of wastewater, wastes, and return water is one of the most important factors threatening the environment and the quality of water resources in the country. There are many places (especially big cities) where due to lack of wastewater network systems and also overconsumption, the pollution problem restricts the available water resources. Table 4.4 shows the main indices for the pollution of national water resources for the year 2001 and the present status indicating a rapidly deteriorating condition (MoE 2016).

4.4.5 Desalination

The Iranian government envisages massive investments in seawater desalination and in the required water conveyance infrastructure, mainly from the Southern shores of the country. Initially, desalination plants are to be built to supply coastal cities, while at the second stage, cities in the central plateau are to be served as well. The plants and pipelines are expected to be financed by the private sector under Build-Own-Operate (BOO) Contracts where the government pays annual fees for the water produced. Such contracts for desalination plants already exist on a small scale with Iranian companies and are expected to be extended to larger contracts

Table 4.4 Main indices for the pollution of national water resources

Indicator	Year 2001 (BCM)	Present status (BCM)	Average annual growth (%)
Urban and industrial wastewater	3	8	4
Total wastewater and return water	29	40	1.3

with international companies. The power for the desalination plants is expected to be provided, at least partly, by a "small" nuclear power plant. At some stage, the authorities claimed that 45 million people in 17 provinces would be the target for the scheme through 50 desalination plants, without specifying the costs or funding sources and/or the environmental implications. To this one may add the huge potentials for desalination of sizable brackish water resources inland.

4.4.6 Water and Wastewater Treatment

Treatment for drinking water production involves the removal of contaminants from raw water to produce water that is pure enough for human consumption without any short-term or long-term risk of any adverse health effects. Substances that are removed during the process of drinking water treatment include suspended solids, bacteria, algae, viruses, fungi, and minerals such as iron and manganese. The processes involved in removing the contaminants include physical processes such as settling and filtration, chemical processes such as disinfection and coagulation, and biological processes such as slow sand filtration. All these processes are managed and mastered at water authorities in Iran, where currently 190 water treatment plants spread throughout the country are in operation with five major ones in Tehran catering for a population of some 12 million people.

Treated wastewater is not widely reused in the urban sector, perhaps more because of cultural impediments and lack of acceptance by the general public. Wastewater is, however, extensively treated more often than not, locally. In recent years the municipalities have adopted the use of treated wastewater for urban services such as green development and maintenance.

4.4.7 Urban Water Pricing

The price of drinking water in Iran is very low compared to many European countries. Despite attempts to normalize drinking water prices steadily during various national development plans, public welfare considerations and external pressures, lately sanctions, has hindered efforts on this front. Minimum tariffs are observed, hardly covering the cost of water distribution.

4.4.8 Public/Private Sector Partnership and Water Utility Management

Unlike the energy sector, perceived to be more related to the industrial perspective, the urban water and wastewater sector is generally considered to be part of the public services and heavily influenced by government budgeting and planning programs. This has greatly hindered private sector involvement as encouragements and incentives have been limited. Efforts are being made by the authorities to define and design public/private partnerships for water utility management at local scale (MoE 2016).

4.5 Water and the Environment

Allocations on environmental flows are generally planned at all study levels of infrastructural developments. Unfortunately more often than not they are not fully observed and adhered to, given the harsh conditions prevailing, socially and economically, as a result of severe water scarcity. Immediate and ad hoc measures are usually adopted for lakes and wetlands during drought periods. This is primarily achieved by managing outflows from water-regulating bodies such as upstream reservoir dams and restricting water abstractions for agricultural/industrial use marginally.

4.5.1 Water Fauna and Flora

Since Iran enjoys a variety of climate patterns with the Caspian Sea in the North, the Persian Gulf, and the Sea of Oman in the South plus two huge mountain ranges encompassing a vast Sahara, the diversity of fauna and flora in the land is large and somewhat unique. The green coverage in the north of the country is steady and dense. The total forest area in the north is estimated to be 1.8 million hectares. The species variability is quite wide in this region, ranging from variety of vertebrates to general/special invertebrates. The Caspian Sea alone is home to more than 850 variety of species with the Sturgeon as the shining jewel, producing one of the best Caviar in the world.

The mountainous region of Iran has its unique fauna and flora with sparse forest land with more than 10 million hectares of Oak forests. Again the species variety in this region is quite varied.

The desert zone has a unique biodiversity and the communities inhabiting the perimeters of the great central deserts of Iran are among some of the most ancient and rich civilizations in the world mastering the art of harnessing the harsh environment throughout their long history. Isfahan and Shiraz are among the famous cities in Iran's central plateau, which have shown great resiliency in preserving the fauna and flora in the face of their historical development. The adverse effects of rapid pace of development alongside the looming impacts of climate change have added to the worries of instability and endanger of the fauna and flora in the country. The annual Red List published by the IUCN indicates the situation in various categories of critical, vulnerable, and in danger of extinction. Table 4.5 shows a sample IUCN survey for vertebrates in Iran.

Table 4.5 The situation of vertebrate species of Iran based on IUCN's (International Union for Conservation of Nature) Red List, 2009

Type of vertebrate	Number of species worldwide	Number of endangered species	Percentage of endangered worldwide	Number of species in Iran	Number of endangered in Iran	Percentage of endangered in Iran
Mammal	5488	1141	21	194	17	8.4
Bird	9990	1222	12	521	20	3.8
Crawling	8734	423	5	216	9	4.2
Amphibia	6347	1905	30	20	4	20
Fish	30,700	1375	4	180	28	4.5
Total	61,259	5966	9.7	1131	78	6.8

4.6 Energy–Water Nexus

Water and energy are coupled in intimate ways. Many of our technical processes of harnessing, extracting, and producing energy utilize water. Similarly, water extraction, treatment, distribution, and disposal processes consume energy. This interdependency, often referred to as the "water–energy nexus," has been increasingly highlighted as an important issue for future planning and strategic policy considerations. Historically, the per capita water resources in many countries were abundant and the economic sensitivities to variations in natural systems were low (or at least not well understood). Water and energy systems were largely treated independently. In recent years, however, with rapid population growth, and increasing awareness of impending (potentially significant) changes in regional climate and water cycle, there has been a growing need for integrating the planning and design of energy and water systems (Webber 2008). Iran is adopting this approach steadily and primarily in academic and research institutions. Further capacity development on the subject is being sought by respective government bodies.

For energy availability and consumption, the electricity generation and consumption and oil production and consumption were considered. Iran has sufficient (in fact surplus) capacity to meet its own electrical power needs. This is partly due to the fact that consumption is not constant over time, and different seasons and times of the day require larger installed capacity. Data on planned capacity indicate that over the next 5 years electric capacity will increase by 31% on average (MoE 2016).

In order to appreciate the mutual dependencies of water and energy systems in Iran, water consumption in energy production and energy consumption in water systems may be quantified by evaluation of the water intensity in energy production and energy intensity in the water value chain. The high dependence of the installed electric capacity on freshwater could constrain electricity production if freshwater levels significantly decrease because of climate change. From a resource standpoint there is wide variation in supplies of freshwater and energy resources. There are, however, some common trends in Iran, such as increasing urbanization, and decreasing share of agriculture in national GDPs. These trends can have important consequences for water (and indirectly energy) demands.

In Iran, the main segments of water use along the energy value chain are fossil fuel extraction and refining and electricity generation. Water consumption in oil extraction is much lower than what is consumed through evaporation in cooling processes in power plants. Iran is one of the MENA region countries which are large producers of oil and petroleum products, so the collective effects on water consumption can be significant (Siddiqi and Laura Diaz 2011).

4.7 Special Issues

Iran has shared waters and boundary rivers with all of its neighbors. Iran has all kind of transboundary surface and groundwater. Some transboundary rivers flow to Iran; for example, Aras Transboundary River flow enters to North West of Iran from Caucasus Transboundary Basin. Harirud and Hirmand (Helmand) Transboundary Rivers enter to eastern parts of Iran from Afghanistan.

4.7.1 Conflicts, Negotiations, and Agreements on Shared Water

Iran and its neighbors have several transboundary rivers and shared wetlands and for which many agreements and protocols are in place to help them avoid disputes and also provide new space for more cooperation. Figure 4.5 depicts transboundary river basins of Iran.

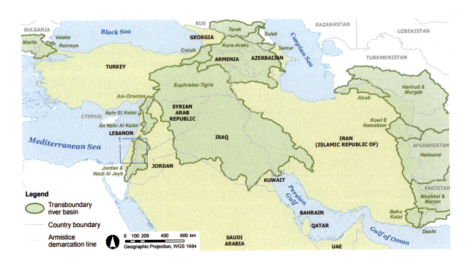

Fig. 4.5 Transboundary river basins

4 Iran's Water Policy

Table 4.6 Iran's important transboundary rivers

River	In/out	Country's position in transboundary basins			Treaty
		Upstream	Downstream	Boundary river	
Aras	In	Turkey, Armenia	Iran, Azerbaijan	Iran, Armenia, Azerbaijan	Yes
Sari Su and Ghare Su		Turkey	Iran	–	Yes
Harirud		–	–	Iran – Turkmenistan	Yes
Hirmand (Helmand)		Afghanistan	Iran	–	Yes
Astarachai	Out	–	–	Iran – Azerbaijan	Yes
Atrak		Iran, Turkmenistan	Turkmenistan	Iran – Turkmenistan	Yes
Nihing		Iran	Pakistan	–	No
Northern Khorasan Rivers		Iran	Turkmenistan	–	Yes
Western Boundary Rivers		Iran	Iraq	Iran – Iraq	Yes

Table prepared by Transboundary Rivers and Shared Water Resources Bureau

Table 4.6 shows some of the most important Iranian transboundary rivers on most of which there are some treaties, agreements, and protocols between Iran and neighboring countries on issues pertaining to water rights, hydropower, navigation, water quality, and so on.

Iran and neighboring countries have wide ranges of cooperation with legal, institutional, technical, and management dimensions and also have joint monitoring committees for discharge measurements and river pollution, common water works, and so on.

Iran and neighboring countries have experienced many types of cooperation and challenges. For example, construction of Aras Dam over transboundary Aras River (between Iran and Azerbaijan (Former Soviet Union)), Khoda Afarin Dam over Aras Transboundary River (between Iran and Azerbaijan), or Doosti (friendship) Dam over Harirud Transboudary River (between Iran and Turkmenistan) shows the institutional, technical, and management capacities of Iran and neighboring countries for development of benefits sharing in their transboundary rivers and also avoiding new challenges or disputes in the transboundary basins.

Despite the improvement on transboundary waters cooperation, some important challenges still remain. Table 4.7 shows some of these challenges between Iran and each neighboring country.

Iran has traditionally engaged in cooperation with international bodies, and some of the most recent ones are listed in Table 4.8.

Table 4.7 Challenges in transboundary waters between Iran and neighbors

Country	Challenges
Afghanistan	Weak water resources management in the Helmand (Hirmand) basin
	Development of diversion dams and canals without attention to downstream required waters
	Desiccation of Hamoon wetlands
	Development of sand storms and wind erosion in lower Helmand (Hirmand) basin
	Opium cultivation
	Lack of agreements on Transboundary Harirud River basin between Iran, Afghanistan, and Turkmenistan
	Low water efficiency in the agricultural sector
Armenia	Aras River pollution
	Copper-mining factories and tailing dams discharge to Aras River
	Measurement stations
Azerbaijan	Aras River flood control measures resulting in bank erosions on the Iranian side
Iraq	Demise of Mesopotamia marshes due to GAP project in Turkey, and instigating dust storms in western and central parts of Iran
	Lack of agreements on Transboundary Aquifers
	Low water efficiency in agriculture sector
Pakistan	Lack of agreements on Nihing Transboundary River basin and Mashkil Transboundary Aquifer
Turkey	Lacking any agreement for transboundary Aras River
Turkmenistan	Need for reconstruction of Qare Qum Canal for water-loss prevention
	Weak water resources management
	Low water efficiency in agriculture sector

Table 4.8 Iran and international organizations cooperation on transboundary waters since 2005

Title of cooperation	International organizations	Year of cooperation
Rehabilitation of Hamoon Transboundary Wetland and its unique lakes shared between Iran and Afghanistan at the downstream of Helmand (Hirmand) River in Sistan Delta for implementation of IWRM in the Helmand (Hirmand Transboundary Basin)	UNEP, UNDP, and GEF	2005–2007
Reduction of Aras Transboundary River Degradation	UNDP and GEF	2005–2007
Making the most of Afghanistan's river basins opportunities for regional cooperation	EWI	2009–2010
Second assessment of transboundary rivers, lakes, and groundwaters (as focal point of Caucasus and Central Asia)	UNECE	2009–2012
ET (Euphrates and Tigris) benefit sharing project	SIWI	2010–2012

4.8 Conclusions

Iran, today, is facing a water deficit in general and moderate to severe scarcity in many of its water-related development and maintenance operations. Precipitation has traditionally been low and in average about one third of the global average precipitation with an uneven spatial and temporal distribution. The country is located in the arid and semiarid regions of the world and although Iranians have been traditionally harnessing water resources ingeniously and efficiently, over the last four decades, the rapid pace of development, demographic expansions and population growth, urbanization, disintegrated large agricultural expansions, climate change and variation, and perhaps most importantly mismanagement of water resources have taken their toll on a barely stable water resources supply/demand chain. To these challenges, one must add the ensuing environmental implications and direct impacts on water quality issues. Social, economic, and political complications may stem from the worsening water scarcity scenarios. For its multifaceted nature and multisectoral role, the need has arisen for a paradigm shift in water resources planning and management. For this, perhaps institutional frameworks should be redesigned to accommodate as much public participation and capacity development at all levels should rise to the top of the agenda. A good deal of research is being conducted at research and academic institutions; however, a concerted effort is required to integrate actions based on pragmatic visions and perspectives.

References

Country Report. (2014). Ministry of Jihad Agriculture (MoJA), IR Iran.
Country Report. (2016). *Water resources status and plans*, Office of water resources planning, Ministry of Energy (MoE), IR Iran.
Iran National Census Report. (2016). Iran's National Statistical Centre.
Karamidehkordi, E. (2010). Country report: Challenges facing Iranian agriculture and natural resource management in the twenty-first century. *Human Ecology, 38*, 295. https://doi.org/10.1007/s10745-010-9309-3.
Siddiqi, A., & Laura Diaz, A. (2011). The water–energy nexus in Middle East and North Africa. *Energy Policy, 39*(8), 4529–4540.
Webber, M. (2008, October). *Energy versus water: Solving both crises together*. Scientific American, Special Edition.
Yazdandoost, F. (2016). Dams drought and Water shortage in Todays Iran. *Iranian Studies, 49*(6), 1017–1028. https://doi.org/10.1080/00210862.2016.1241626.

Dr. Farhad Yazdandoost received his university education in the UK, where he gained engineering and consulting experience. Since joining the K. N. Toosi University of Technology, Tehran, in 1991, he has served as its Dean of the Faculty of Civil Engineering and Vice-Chancellor for Research and is currently serving as Vice-Chancellor for Global Strategies and International Affairs. While seconded to the Ministry of Energy, he was appointed as the Founding President of

Iran's National Water Research Institute, directing research, innovation, and capacity development. Commissioning numerous national and regional research and development projects is some of his achievements which he later pursued in his capacity as the Director of the Regional Centre on Urban Water Management, under the auspices of the UNESCO.

He has authored many scientific and technical publications and has led national/regional research programs and projects. He has supervised over 100 postgraduate (PhD and Master) students, and his research interests include resilience and risk management and integrated approaches to water resources management and engineering. The development of toolboxes addressing challenging water management problems and advancement of resilient approaches to various water engineering issues has earned him insight into providing practical solutions to complex problems. He has held leading roles in international learned and scientific associations and has contributed widely as Coeditor of scientific journals.

Chapter 5
Water Policy in Jordan

Tala Qtaishat

Abstract Jordan's water resources are estimated at a long-term average value of 8191 MCM/year. Groundwater accounts for about 61% of water supply. Water scarcity threatens Jordan's development. About 93.5% of the country receives less than 200 mm of rainfall, and only 0.7% of the country has annual precipitation of more than 500 mm. Drought-occurrence periods are one of the most serious factors affecting water supply. The quality of treated effluent allowed to be discharged into Wadies follows Jordanian Standards. Exploitation of the aquifers increases the salinity level. One of the main weak institutional performances in Jordan's water sector is the overlapping responsibilities between the Ministry of Water and Irrigation (MWI), with Water Authority of Jordan (WAJ), and Jordan Rift Valley Authority (JRVA). Renewable water supply currently only meets about half of total water consumption. This is caused by unsustainable groundwater extraction, including thousands of illegal private wells. Despite the huge investments in the water sector programmed through the year 2025 the future food and water security is under threat unless the government strategy is fully implemented. The water deficit will grow from about 160 Mm^3 in 2015 to 490 Mm^3 by 2025. Without new water sources, only 90 m^3 per capita per year will be available by 2025. The Disi water conveyance project and the Red Sea-Dead Sea Canal will help increase the supply, but will not be sufficient to satisfy long-term demand. A combination of reduced demand and rationing distribution programs for domestic uses as well as the re-use of wastewater flows for irrigated agriculture will likely help bridge the gap.

Keywords Jordan · Water policy · Water resources · Water users

---

T. Qtaishat (✉)
The University of Jordan, Amman, Jordan
e-mail: t.qtaishat@ju.edu.jo

© Springer Nature Switzerland AG 2020
S. Zekri (ed.), *Water Policies in MENA Countries*, Global Issues in Water Policy 23,
https://doi.org/10.1007/978-3-030-29274-4_5

5.1 Introduction

Jordan is an arid to semi-arid country with total population of 9.5 million; the number of Jordanians is around 6.6 million while the number of non-Jordanians who reside in the country is around 2.9 million, representing 30.6% of the overall population (Department of Statistics [DOS] 2017). The country has a total land area of 88,780 km^2; about one-third is dry land while the other two-thirds are irrigated land (World Bank 2016). Jordan is considered to be a water-poor country due to the water-supply shortage. Water scarcity threatens Jordan's development. Precipitation is very low, and it ranges from 30 to 600 mm annually. About 93.5% of the country has less than 200 mm of rainfall, and only 0.7% of the country has annual precipitation of more than 500 mm (UNESCO 2012). The variable and low rainfall with the high evaporation rate and droughts all contribute to low water-resource reliability and availability.

Jordan's water resources depend mainly on rainfall which is estimated at a long-term average value of 8191 MCM/year (MWI 2015a). Groundwater accounts for about 54% of Jordan's water supply; this water comes from 12 groundwater basins (Fig. 5.1). Jordan's groundwater is of two types, renewable and fossil. The latter constitutes 5% of the total groundwater storage. The safe-yield abstraction quantity from renewable groundwater is 275 MCM annually (MWI 2015a). For the fossil-water quantities, the safe-yield abstraction from groundwater for 50 years is about 143 MCM annually (MWI 2015a). The pumping for renewable groundwater was about 160 MCM in 2015 (MWI 2015a). The groundwater's quality varies from one aquifer to another; salinity ranges from 170 to 3000 ppm (Hussein et al. 2005).

Surface-water resources have two principal parts: base flow and flood flow. Base flow is derived from groundwater drainage through springs. Surface water is about 37% of the total water supply and develops through 15 water basins which are distributed across the country (Fig. 5.2). The main surface resource is the Yarmouk Basin in the north which contributes almost 50% of Jordan's base- and flood-flow waters (Al-Ansari et al 2014). In 2015, the actual supply of surface water was 730 MCM/year (MWI 2015a). Other water-supply sources are brackish water and treated wastewater. In 2015, the treated wastewater was 147 MCM/year (MWI 2015a).

The per-capita water availability dropped from 3600 m^3 in 1946 to 145 m^3 in 2008 (UNESCO 2012).

5.2 Administrative, Legal, and Institutional Aspects of the Water Sector

In terms of water governance, the Ministry of Water and Irrigation (MWI) is responsible for the overall strategic direction and planning; the MWI works with the Water Authority of Jordan (WAJ) and the Jordan Valley Authority (JVA) (Fig. 5.3). The

5 Water Policy in Jordan

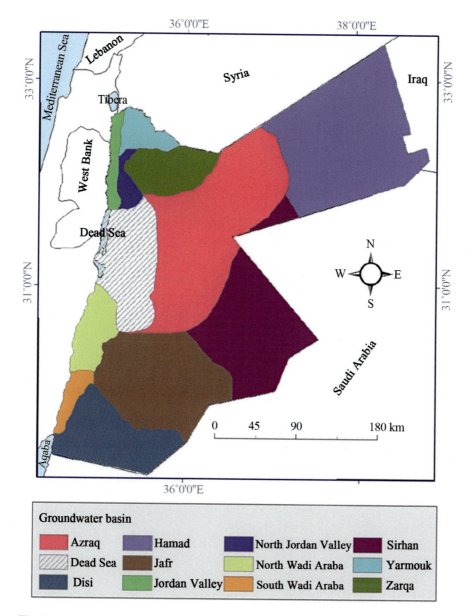

Fig. 5.1 Groundwater basins in Jordan. (Source: Al-Ansari et al. 2014)

National Water Strategy has been created to manage the water sector and to ensure optimal service levels. The National Water Strategy (2016–2025) is part of the MWI's plan which explores the need for more development of water legislation, including the need for a comprehensive water law and moving toward realizing humans' rights to water and sanitation while recognizing these rights and their

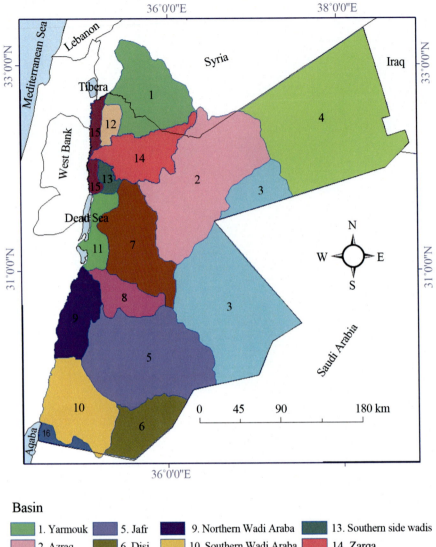

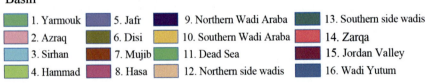

Fig. 5.2 Jordan's surface-water basins. (Source: Al-Ansari et al. 2014)

standard content for all. The strategy's key areas are as follows: (i) Integrated Water Resources Management (IWRM); (ii) water, sewage, and sanitation services; (iii) water for irrigation, energy, and other uses; (iv) institutional reform; and (v) sector information management and monitoring. The strategy also addresses the issues related to climate-change adaptation, transboundary/shared water resources, public/

5 Water Policy in Jordan

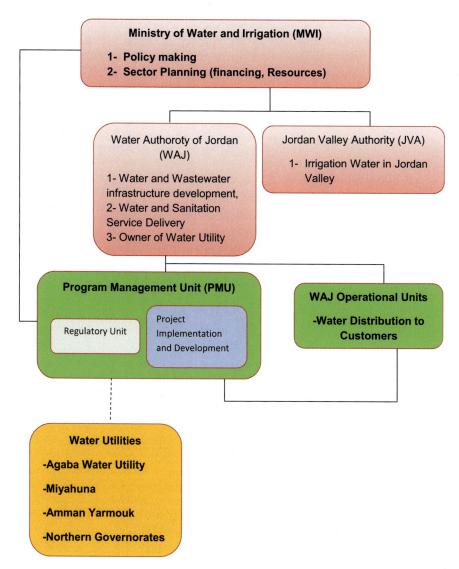

Fig. 5.3 The Different Institutions' interactions and hierarchy in water sector in Jordan. (Source: MWI 2016c)

private partnerships, and the water's economic dimensions. Within the timeframe for this strategy, the MWI aims to adopt a sector-wide, integrated water-resource planning and management approach; develop sector policies and legislation to enhance performance, equitable service provision, and optimization of available resources; initiate institutional reforms to restructure sector management; enhance fiscal discipline for cost recovery; improve internal efficiencies for sector coordination and management; and build technical capacity (MWI 2016a).

Institutional reform is set for greater efficiency and effectiveness, and improve inter-sectoral linkages to generate greater synergy and impact on the health and economic wellbeing of all Jordanians, where each operating agency/unit that is part of MWI would have a clear purpose and incentives to perform effectively and efficiently; these actions are aimed to restructure sector management, enhance fiscal discipline in cost recovery, improve internal efficiencies in sector coordination and management, and build technical capacity.

5.3 Agricultural Water

Agriculture is currently the largest water user. While farmers irrigate less than 10% of the total agriculture land, the agricultural water is estimated to be 700 MCM (Fig. 5.4). The agricultural sector contributed about 3–4% to the gross domestic product (GDP) in 2013 (MWI 2016a, b). The Jordan 2025 vision calls for an increased agriculture GDP share. Although agriculture comprises a relatively small share of the GDP, it provides most of the agricultural production and offers the high percentage of direct agricultural jobs. Cropping patterns, water-reallocation policies, and different irrigation technologies will be adopted by 2025, resulting in yield gains and water savings. Figure 5.4 provides an overview of the agricultural sector's water use in Jordan.

In 2015, there were 52 private desalination plants operated by farmers to desalinate brackish water for irrigation purposes and desalinate about 10 MCM annually.

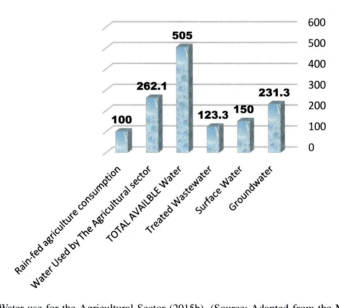

Fig. 5.4 Water use for the Agricultural Sector (2015b). (Source: Adapted from the Ministry of Water and Irrigation MWI 2015b)

Brackish water with salinity between 2000 and 8000 ppm is pumped from wells at depths between 100 and 150 m. The facilities are generally in operation 24 h/d in summer and 8 h/d in winter. The only energy source used to run the plants is electric power.

The surface, groundwater, and wastewater are located in this section because the discussion covers only their use in agriculture, since agriculture is main user of available water in Jordan.

5.3.1 Surface Water

Surface water provides about 27.2% of Jordan total water supply. Jordan developed surface-water resources where they produced about 259 MCM in 2014 and are projected to be about 329 MCM by 2025 mainly through water harvesting by building small dams on the Wadies (MWI 2016a, b). The Jordan and Yarmouk Rivers, major sources of surface water, now provide less than 25% of the shared water that flows in the Yarmouk River Basin, amounting to about one-third of the proposed share to distribute water among the riparian countries (MWI 2016a, b). Rainfall is seasonal, localized, and unpredictable. High evapotranspiration rates diminish the available water's value.

5.3.2 Groundwater

Groundwater contributes about 59.6% of the total water supply. Out of the 12 major groundwater basins, 6 are over-extracted; 4 are at capacity; and 2 are underexploited (MWI 2016a, b). Increasing the overall water extraction to meet national needs carries a high cost; Jordan is now accessing nonrenewable water resources from fossilized deep-water aquifers. Groundwater from the nonrenewable Disi Aquifer (about 100 MCM/in year 2015) contributes to Jordan's water supply for domestic and agricultural needs (MWI 2016a, b).

Jordan's groundwater is in renewable and nonrenewable forms in 12 distinct basins. The installed pumps are capable of extracting more water than the safe yield for each basin. There are 11 renewable groundwater reservoirs in the country. Their sustainable yields vary from one aquifer to another, and their combined yield is 275 MCM per year. The average annual abstraction from all basins exceeds the renewable recharge average and currently stands at 159% of that average (MWI 2016a, b). Decisions were made to treat the situation. These measures include: (1) The agricultural sector's share of groundwater resources shall be capped in favor of other sectors with higher economic return per cubic meter. (2) Treated wastewater of quality meeting national standards and complying with public health requirements shall be increasingly used to replace fresh water resources. (3) The government could close the wells in owned lands after compensating their owners' water rights or land value in a buy-back scheme. (4) Legislations pertaining to groundwater management are enforced equally on all well-owners. Strict measures that deter future violations shall be designed and enforced. (5) Groundwater management action plans developed for Azraq Basin and Yarmouk Basin, with the participation of the local community and water users, as well as the one in the Jordan Valley, shall be implemented.

There are extensive nonrenewable reservoirs in the sandstone formations. These reservoirs' water quality varies and is known to be fresh in the Disi-Mudawwara area. Using fresh fossil water from the nonrenewable reservoir in Disi-Mudawwara started in the early 1980s for municipal and industrial purposes in the city of Aqaba. This was followed by utilizing the same aquifer (Disi) for agricultural purposes. Future use of this aquifer is earmarked for the city of Amman's municipal purposes, and pumping for agricultural purposes is being reduced (MWI 2016a, b).

5.3.3 Treated Wastewater

Treated wastewater generated in Amman and Zarqa is a main water resource in the Jordan Valley. It is treated at As Samra Wastewater Treatment Plant (WWTP) in Zarqa governorate and discharged to Zarqa River, which ends in KTD. The flow from the As Samra WWTP to Zarqa River increased from about 61 MCM in 2007 to about 106 MCM in 2015 (MWI 2016a, b). The treated wastewater constituted about 13.2% of water supplied in Jordan in 2015.

The Water authority of Jordan (WAJ) operates 31 wastewater facilities, serving about 6.7 million people. The highest coverage of the wastewater stations is in Amman governorate (about 84% of the population), Aqaba (about 72%), and Jerash (about 69%), while the lowest covering rates are Tafilah (about 31% of its population), Karak (about 20%), and Mafraq (about 8%) (International Resources Group (IRG) 2013). Most of the WWTPs are applying the activated sludge system as standard treatment process. There are some challenges regarding operation and maintenance (O&M) of wastewater treatment plants; they include lack of professional expertise in WWTP process control, inadequate O&M budgets, and lack of performance-based bonus/malus schemes and at the end an effective monitoring system to control and steer the WWTP operations in real-time mode (MWI 2016c).

5.3.3.1 Aquifer Depletion

Modern technology to access groundwater was introduced to Jordan in the late 1950s; legislation to regulate the exploitation of groundwater resources and to have it supervised by the government was introduced in 2002. The difference between the amount of renewable water and the amount extracted slowly led to depletion and salinization of the country's groundwater resources (El-Naqa and Al-Shayeb 2009). The depletion of groundwater aquifers was declared by the MWI's 1998 National

Groundwater Management Policy as the major problem facing Jordan's water sector. The MWI established a law that prohibited drilling new wells in most parts of the country where aquifers are afflicted by depletion and quality degradation.

The policy set specific objectives and principles for groundwater use and management. The groundwater management policy addressed the management of groundwater resources, covering development, protection, and reducing abstraction from each renewable aquifer to sustainable rates. Other measures included continuous enhancement of the groundwater quantity and quality monitoring networks, substitution of fresh groundwater with marginal water in agriculture (brackish, treated wastewater), and adopting strategies for the reduction of groundwater abstractions, to reach the safe-yield levels of 275 MCM/year by the year 2020 within the scope of Water Master Planning, which is the authorized source for data on water monitoring, management, and planning for external users like research institutions or international donors.

Table 5.1 summarizes the measures of groundwater management in Jordan under four levels: reducing groundwater consumption, promoting recharge and retaining groundwater, and regulating groundwater development.

5.3.3.2 Property Rights

The following rules were set by the groundwater policy implemented by MWI to institutionalize the priorities of water allocation: (1) Priority of allocation of groundwater shall be given to municipal and industrial uses, to educational institutes and to tourism. These purposes are deemed to have the higher returns in economic and social terms. (2) Priority shall also be given to the sustainability of existing irrigated agriculture where high capital investment had been made. In particular, plantations irrigated from groundwater shall continue to receive an amount sufficient for their sustainability with the use of advanced irrigation methods. (3) Expropriation of use rights arising from legal use of groundwater or of water rights established on springs rising from groundwater reservoirs shall not be made without clear higher priority need and against fair compensation. (4) Priority shall be given to the use in irrigated agriculture of the reservoirs whose water quality does not qualify them for use in municipal and industrial purposes. (5) Priority for use in agriculture shall also be given to the cases where supplementary irrigation from the groundwater reservoir is possible. (6) A contingency plan shall be made and updated for the purpose of allocating the water from privately operated wells for use in the municipal networks (MWI 2017).

In the Jordan Basin, the magnitude and frequency of conflict increase during periods of drought. During the drought of 1998–99, Israeli and Jordanian members of the JWC brokered a temporary arrangement to modify allocations to reflect water availability, thus resolving the conflict. While the absence of a drought provision in the Treaty of 1994 left Jordan and Israel vulnerable to conflict, the treaty did establish the Jordan Water Committee to resolve conflicts without making permanent amendments to the original agreement (Odom and Wolf 2011).

Table 5.1 Measures of groundwater management in Jordan

Level	Features
Reducing groundwater consumption	Promoting higher water productivity in agriculture:
	Changing cropping pattern, varieties, and agronomic practices
	Micro-irrigation and root-zone irrigation
	Improved water conveyance
	Increasing water-holding capacity of soil
	Reducing waste in processing and marketing
	Reducing urban groundwater use
	Leakage detection
	Reducing domestic water loss and use
	Urban landscaping
	Use of economic incentives
	User fees
	Pricing of energy supplies
	Redirecting subsidies from water-intensive crops
	Water-efficiency incentives
	Smart-card-controlled abstraction and quotas
Promoting recharge and retention of groundwater	Interception and retaining surface runoff and floods
	Field bunding and terracing
	Contour buds and gullis
	Seepage wells and maintaining natural pits
	Injection wells
	Water harvesting from roads
	Recharge bonds, dams, and sand dams
	Flood water retention
	Improved infiltration capacity of land surface
	Permeable urban surface
	Penetration of day Layers
	Increasing infiltration by burrowing action of animals
	Sand dams
	Retaining surface flows
	Gully plugging of drainage canals
	Subsurface dams
	Conjunctive management of surface water and groundwater
	Adjusting surface water delivery to recharge and reuse potential
	Storage of seasonal excess water

(continued)

Table 5.1 (continued)

Level	Features
Regulating groundwater development	Promoting self-regulation
	Enabling laws
	Developing and applying local rules
	Participatory monitoring and assessment
	Joint crop planning
	Local investment in recharge
	Well licensing and well regulation
	Geographic bans
	Licensing of drilling rigs
	Tracking of drilling rigs

Source: Smith et al. (2016)

5.3.4 Reuse of Treated Wastewater

5.3.4.1 Food Safety/Vegetables/Heavy Metals

When using treated wastewater, food safety is considered an important issue in Jordan because it affects the fruits-and-vegetables export sector. In 1990, the Gulf Cooperative Council States (GCC) stopped importing fruits and vegetables because the crops were irrigated by blended fresh water with treated wastewater (Albakkar 2014). Wastewater in Jordan is considered as a main source of irrigation. Crops irrigated with wastewater or blended water are monitored in Jordan. Water quality monitoring in Jordan is covered by the following intuitions: Royal Scientific society, Water authority of Jordan, Jordan Valley Authority, Ministry of Water and Irrigation, Ministry of Environment, Ministry of Health, Meyahouna, Aqaba Water Authority, and North Governorates water Agency (Saidam 2009). The monitoring is conducted through continuous, automated, on-site sampling and analysis, data acquisition, storage and dissemination in one system in real time. As for fruits and vegetables ready for export, they are inspected for chemical residues in specialized laboratory in the Ministry of Agriculture.

The government role in this sector is regulatory and supervisory while encouraging the private operation and maintenance of utilities. The wastewater-treatment plant owners' responsibility is to consider the conformity of the standards with the end user. Jordan follows ISO and GlobalGap standards to monitor use of treated wastewater for fresh fruits and vegetables (Seder and Abdel-Jabbar 2011). In contrast, toxic-material discharge to sewers and sludge use are regulated in Jordan by MWI. The successful utilization of recycled water within Jordan has been made possible by the development and evolution of a sound legislative and legal foundation. There are several sets of standards that have paved the way Jordanian Standards

JS893/95, JS202/91, JS 1145/96, WAJ's regulations for the quality of industrial wastewater to be connected to the collection system, and WAJ's specifications for sewerage works, have been, thus far, the benchmarks against which plans and specifications of treatment plants and wastewater reuse were evaluated. They were established to bring about relative uniformity throughout the country. However, there is a risk in using industrial wastewater which comes from inadequate industrial and municipal wastewater-treatment capacities, and the industrial plants are built near or immediately upstream from potable water supplies (Haddadin and Tarawneh 2007).

5.3.5 Pricing

Appropriate water pricing can be used for optimizing cropping patterns and water distribution, which can also substantially increase production and yields (Table 5.1).

5.4 Agricultural Production

5.4.1 Cost Recovery/Maintenance of the Irrigation Schemes

Jordan's water subsidy is considered to be high because the irrigation water price level is much lower than the total cost recovery which covers the direct and financing costs (USAID 2012). Analyzing the cost of Jordan's irrigation water shows the need for significant price increases to strengthen its financial sustainability. Depending on the level of cost recovery, the minimum price increase that is required for irrigation water could be very large. If the government wants to pursue its objective, as stated in the Government of Jordan's Water Strategy (2009b), that depreciation should also be covered, the irrigation-water price would have to be increased to between JD 0.132 and JD 0.215 per cubic meter, depending on whether billing and collection inefficiencies improve (Van Den Berg et al. 2016). In 2012, the Cabinet of Ministers approved a new pricing policy on irrigation water – even on amounts already granted in existing licenses – with a block price system, where charges increase in relation to the amounts of water extracted (Table 5.2).

Table 5.2 Volumetric prices of water abstracted from replacement wells

Water quantity, (m^3)/a	Water price, JD/ (m^3)	Water price, US$ /(m^3)
1. 0–75 thousand	Free	Free
2. 75–200 thousand	10 Fils/m^3	0.0141
3. > 200 thousand	100 Fils/m^3	0.141

Source: Al-Karablieh and Salman (2016)

5.4.2 Institution Setting of Irrigation Organizations and Farmers' Participation

Recognizing the difficulty of managing farm-level irrigation in Jordan, the government started forming water user associations (WUAs) since 2001 (USAID 2013). Most WUAs are legally registered as independent cooperatives, and they work under government control.

WUAs in the Jordan Valley are classified into three progressive levels in terms of their status: (1) Water councils: They are based on the traditional mechanism of problem solving. Water councils are recognized by the JVA. Each council would have 15–20 elected farmers chosen through prior informal discussion with the concerned farmers. The government is represented through the sub-governor (Al Mutassarif) in the water council. Thus, the council has an executive power. Al Mutassarif may even chair the council. (2) Water user committees: They are also based on or similar to the traditional form of farmers' management that existed before the formation of JVA. A water user associations committee is a group of representatives of farmers elected by the farmers in a general assembly after several informal meetings. Although the associations have no legal status as such, they are recognized by JVA; normally, a letter is issued by the JVA secretary general in this respect. (3) Water user cooperatives: They are the type of associations that have a legal status. Cooperatives follow the Cooperation Law No. 18/1997 and thus they are affiliated to the Jordan Cooperative Corporation (JCC). Cooperatives have their internal regulatory system that specifies the objectives, capital, membership procedure, and financial and administrative issues.

Farmers' participation resulted in direct savings of water resources. A good example is given in Al Kafrein area, where the community was able to optimize the irrigation scheme, reduce leakages and illegal connections to the network, and thus reduce the water released from the dam to the network from 12,000 to 6000 m³/day. This was achieved only in 2 months after handing over the water distribution task to the WUA. The WUA of Al Kafrein also pointed that the proper management of water enabled them to withstand and manage their farms even with less amount of water in the dry seasons.

5.4.3 Irrigation Efficiency

5.4.3.1 Technology Adoption/Subsidy and Other Policies

Irrigation-water efficiency in the Jordan Rift Valley is around 65% while on-farm irrigation efficiency in the highland where drip irrigation from groundwater is utilized is around 85%. The center-pivot system's performance is between 76% and 84% (Shatnawi et al. 2005). Irrigation efficiency in the Al'Azraq area is around 75% for sprinkler and 85% for drip irrigation (Abu-Awwad and Blair 2013).

5.4.3.2 State-of-the-Art Technology/Smart Irrigation and Innovations

In 2010, the SMART II project was initiated in the Lower Jordan Valley for three countries: Jordan, Palestine, and Israel. The project aimed to develop the Integrated Water Resource Management (IWRM) concept to ensure optimized and sustainable use for all the region's water resources in addition to achieving socioeconomic conditions that satisfy the demands of society's diverging groups. The overall goal of IWRM is to ensure that national water-resource management is based on the principles of sustainable use, economic efficiency, social equity, and environmental and ecological sustainabilities.

There are several methods that farmers use to irrigate their lands in the Jordan Valley. Farmers' choices of irrigation technique depend on the kind of crops they grow, their financial reality, and the information they have access to. In the Jordan Valley, 68% of the farmers use drip irrigation, 30% use surface irrigation, and 2% use sprinklers. It is estimated that water savings through optimization of irrigation could reach almost 40 MCM/year by the year 2020 at a cost of 0.5 US\$/m^3 (Klinger et al. 2012).

5.4.3.3 Research

A large amount of research covering different aspects of the water demand was conducted for Jordan's water sector (Al-Ansari et al. 2014; Hjazi 2010; Arabiyat 2005). The Water Demand Management Program ensures further reduction in water use, reduces water losses through the distribution supply net, and prevents pollution. In addition, it helps minimize water disposal in nature, makes efficient use of available water resources, plans for future new water resources prudently, and finally imposes a real cost for water supply that would be acceptable. In addition to the above, public awareness program is to be put in action. Such a program should be used in schools as well as the media. The public is to be aware of the problem and how they can assist with overcoming the water shortage crisis (Al-Ansari et al. 2014).

Another study found that there are a variety of water-use efficiency programs that can be implemented. However, the cost of these programs is usually the major factor that influences the implementation. Past and ongoing experience on WDM programs indicates that strong emphasis should be devoted to retrofitting programs such as aerators and showerheads. The analysis shows that they are most cost effective in the residential sector, government buildings, mosques, and schools. A WDM Code for new buildings should be in place and enforced (Hjazi 2010).

Field studies using groundwater for irrigation showed that water-pricing policies based on volumetric charges could have little impact on water consumption as water demand is inelastic. Irrigation water consumption has decreased significantly only when water prices have exceeded the 0.5 US\$/m^3 (Arabiyat 2005).

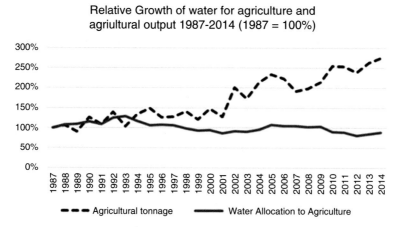

Fig. 5.5 Relative growth of water for agriculture and agricultural output in Jordan during 1987–2014. (Source: Namrouqa 2017a, b)

The Jordan Water Authorities are in the process of increasing the water supply by implementing the Red-Dead Sea canal which will provide desalinized water for agriculture and drinking water. A feasibility study found that this project is financially viable (Blogger 2012).

5.4.4 Food Security vs Virtual Water/Food Imports

The agricultural sector's water use constitutes about 64% of the water demand in Jordan; thus, food security and water are strongly linked (Fig. 5.5).

From Fig. 5.5 it is noticed that agricultural output had increased during 2005–2014, while the water allocation to agriculture had decreased. This was due to the water conservation policies implemented during the same period.

Jordan has established trade and economic agreements both at regional and bilateral levels with around 86 countries from all over the world (specifically, 19 agreements are with Arab countries and 67 with non-Arab countries). To mention are hereby the Greater Arab Free Trade Agreement (GAFTA); the Agadir Agreement, which is creating a free trade area between Jordan, Egypt, Tunisia, and Morocco; the Jordan-United States Free Trade Agreement, which entered into force in 2001; the Euro-Jordanian Association Agreement of 2002; and the FTAs with Singapore, Canada, and Turkey.

Jordan's agricultural and food exports are mainly focused on neighboring countries. Jordan's major agricultural exports are fresh fruits and vegetables, processed meat, food preparations, and live sheep (Tables 5.3 and 5.4).

Table 5.3 Virtual water import (VWI) per major trade partner (Million m³/year) 1996–2005

County	Green VWI	Blue VWI	Gray VWI	Total VWI	Major products
USA	697	88	123	908	Wheat – 66%, maize – 16%, rice – 8%
Syria	626	92	122	840	Barley – 78%, animal products – 4%
Argentina	641	11	31	683	Wheat – 25%, maize – 38%, soybean – 35%
India	434	35	29	498	Animal products – 40%, soybean – 34%, coffee – 7%, wheat – 6%, cotton – 4%
Iraq	172	222	156	550	Barley – 69%, industrial products – 29%
Malaysia	319	0.5	14	333	Oil palm – 97%
Indonesia	238	0.1	17	255	Oil palm – 88%
China	133	22	83	239	Cotton – 71%, industrial products – 14%, animal products – 6%
Turkey	172	21	25	218	Wheat – 41%, barley – 29%, chickpeas – 13%, cotton – 7%
Ukraine	173	4	30	208	Barley – 60%, sunflower seed – 16%, industrial products – 14%, wheat – 9%
Australia	93	41	3	138	Animal products – 53%, rice – 32%, barley – 12%
Total	3698	536.6	633	4870	

Source: Hoekstra and Mekonnen (2012)

Table 5.4 Value of main exports during 1995–2013 (1000 US$)

Product	Year				
	1995	2000	2005	2011	2013
Total merchandise trade	1,772,340	1,899,000	4,301,419	8,006,449	7,910,716
Fruit + vegetables	99,847	105,544	273,247	590,955	724,703
Oranges + tang + clem	9016	3995	6737	12,304	50,393
Oranges	3728	1891	2859	5337	6674
Grapes fresh	1196	1161	2521	1988	1702
Olives	0	10	4500	5200	15,038
Olives preserved	493	372	2071	5555	6770
Lettuce and chicory	1987	4426	7340	7154	11,447
Potatoes	5596	2413	7200	3943	9180
Tomatoes	24,745	34,262	105,707	224,847	316,321
Poultry meat	5383	448	5622	30,178	41,815
Eggs in the shell	3095	5761	4369	10,172	5954

Source: FAOSTAT (2017)

5.4.5 *Water Salinity and Other Pollution Problems (Nutrients and Manure)*

Over-pumping the underground reservoirs increases the salinity level, such as in the North Badia and Jordan Valleys. Over-drafting the surface reservoirs, such as in the Azraq Oasis, and dumping wastewater from the Khirbit Al-Samra treatment plant polluted the Amman-Zarqa Basin. Saline and brackish water resources are available in many places in the country, especially in the South Jordan Valley (Permaculture Research Institute 2005).

5.5 Urban Water

Investments in municipal networks in Jordan remain inadequate. Although the level of services in the water supply in Jordan is fairly high, with service to 97% of the population in the urban areas and 83% in the rural areas, distribution systems are still far from optimal and efficiencies are still low. The unaccounted-for water in the municipal networks was estimated to be 55% of the quantity supplied in 1995. In 2004, over 50% of the water entering the city's distribution system was effectively unaccounted for, with half of this being lost by leakage and the rest due to poor administration and inadequate billing. Although the losses in the urban water supply networks have become a growing concern, water companies have budgetary and other constraints that hinder addressing the problem. According to Trojan and Morais (2012), problems encountered in the maintenance and management of the water-supply systems are indicated by the lack of decision support models that give a manager an overview of the system.

Households pay progressive drinking water and wastewater tariffs. Households are charged for water utility services based on an increasing block tariff, which provides incentives to save water. The polluter pays principle is enforced via sewage charges added to the water bill. Since the sewage charge increases the total water and wastewater bill, it also functions as an incentive to reduce water consumption (MWI 2016b) (Table 5.5).

Table 5.5 Urban water price in Amman in 2016

Water quantity, (m³)/month	Consumption per month in US$/m³		
	15 m³	50 m³	100 m³
Price	0.78	2.98	3.06
Fixed charge	0.45	1.13	0.57
Variable charge	0.22	1.44	2.07
Other charges			
VAT	0.11	0.41	0.42
Total	1.56	5.96	6.12

Source: WAJ Amman, Jordan, 2016

The level and design of water and wastewater tariffs paid by households could not be further improved to approach cost-recovery levels and provide additional incentives to save water. Since households receive an intermittent water supply, their consumption is already limited by the capacity of residential water-storage tanks. The current progressive tariffs were linked to consumption per person. There would be an opportunity to introduce a tariff structure with more targeted incentives to save water. A prerequisite for such a measure is the availability (or development) of good household records on the number of inhabitants/family. There is also an economic barrier related to the affordability of more expensive water. Any change in tariff levels should therefore be gradual and should respect the economic status of households.

Most policymakers in Jordan agree that water for human consumption, including drinking, cooking, bathing, and cleaning, should be given priority over other uses (Qtaishat 2013).

About 48% of the total water resources are used in the industrial and domestic sectors: 4% and 44%, respectively (MWI 2016a, b). Water-reallocation strategies, such as changing the cropping patterns and moving away from crops where the product value per unit of water is relatively low, could improve the overall economy. Water reallocation could have a substantial effect on the municipal and industrial sectors and might lead to an increased GDP for the region, creating jobs in the industrial sector (Qtaishat 2013). A small water reallocation (5%) from agriculture could dramatically increase the water available for other sectors, particularly the municipal sector. Reallocation will be based on the following conditions (MWI 2016a, b): Each governorate shall retain its available water for its sole needs, unless there is severe shortage in other governorates; in this case, it will be transferred to the geographically nearest governorate and to the governorate of highest need, with due consideration to sustainability, long-term feasibility, availability of infrastructure. Shared water resources shall be allocated to the governorate of the highest need and geographically closest, and which can technically receive the water. Supplied water shall be increased to achieve the target shares by the reduction ratios in Non-Renewable Water (NRW). Availability of water infrastructure shall be insured during the reallocation process. Plans shall be made for infrastructures to meet the long-term needs and/or the structure life time.

Among the options for water-supply augmentation are desalinating saline groundwater, brackish drainage water, and seawater. In Jordan, desalination is receiving considerable attention from scientists, resource planners, policymakers, and other stakeholders. In 2010, desalination provided 30 MCM of water in Jordan, and by 2030, desalination is projected to provide about 170 MCM of water (Al-Mutaz 2005; El-Sadek 2010; World Bank 2007; United Nations 2010). Several brackish springs have been identified in various parts of the country. Estimates of stored volumes of brackish groundwater for the major aquifers suggest immense resources, but not all of these quantities will be feasible for utilization.

Currently, there are 44 public desalination plants and an additional 10 under construction that desalinate about 80 MCM annually. All these plants are run or will be run by WAJ to treat saline water for drinking water supply (Al-Karablieh and Salman 2016). Currently, desalination is primarily used for the industrial and tourism sectors because of the high cost of seawater desalination. Utilizing desalination for other purposes (agriculture and municipal) will depend on technological improvements that result in reduced overall and marginal costs (Qtaishat 2012).

The desalination process that will be implemented for the Dead-Red Sea canal is hydropower generation and reverse osmosis desalination facility with a capacity of 850 MCM/year. The estimated cost for desalinization in this process was at US$ 0.46, out of which US$ 0.23 was for electricity, US$ 0.03 for membranes, US$ 0.10 for labor, US$ 0.07 for chemicals, and US$ 0.03 for parts (Ornwipa and Rodriguez 2008). On the other hand, the desalination cost for brackish water was estimated at US$ 0.33 for large plants and US$0.48 for small plants (Qtaishat et al. 2016).

In Jordan, water reuse is an existing tool to manage scarce water resources. Over time, wastewater reuse has changed from simply irrigating field crops with untreated wastewater to a sophisticated reclamation process for agricultural, industrial, and domestic reuse (Durham et al. 2005). The most practical solution for water scarcity is reusing domestic wastewater for some non-potable municipal purposes, such as flushing toilets, irrigating green spaces, and agriculture. Reusing wastewater is cheaper than developing new water supplies and protects existing sources of valuable fresh water from overexploitation (Faruqui 2002). The As-Samra plant is the largest wastewater treatment plant in Jordan, and it can treat about 75% of the 267,000 m^3/day (Ammary 2007). The government buys water from the As-Samra plant for approximately $1.10/m^3$ for irrigation in the Jordan Valley farms (Al-Zu'bi 2007). It should be noticed that this cost is higher than desalinization of brackish water. The effluent Biochemical Oxygen Demand (BOD5), Chemical Oxygen Demand (COD), and Total suspended solids (TSS) from As-Samra WWTP complies with Jordanian standards for reclaimed wastewater discharge to streams, groundwater recharge, irrigation parks, reuse for irrigation of cocked vegetables, fruits, and trees, and for reclaimed wastewater reuse for fodder crops (Myszograj and Qteishat 2011).

The investment and operation costs for wastewater treatment and reuse are high. However, treated wastewater is increasingly being used for agricultural irrigation. Many efforts, such as increasing awareness and information campaigns, are needed to encourage participatory approaches by the farmers.

5.6 Water and Ecosystems

There are two components of ecosystems: (a) biotic components: plants, animals, fish, and parasites and (b) abiotic components: water resources, soil, and climate. Ecosystem regulations, maintenance, and provision are directly affected by the habitat quality and the water quality. A study showed the existence of high pressure on natural habitats in the northern region of the Jordan Valley. In this area, agriculture and industry are the main activities (El-Habbab 2013).

Jordan established several groundwater protection zones, out of which was Wadi Shuaib ecosystem. An economic evaluation of the Wadi Shuaib ecosystem found that it is important to implement groundwater protection zones; the main benefits from these practices are to protect drinking water and the ecosystem as well as to

Table 5.6 Tariff structure for wastewater in Amman in 2016

Wastewater quantity (m³)	Consumption per month in US\$/m³		
	15 m³	50 m³	100 m³
Price	0.15	0.99	1.40
Fixed charge	0.02	0.01	0.00
Variable charge	0.11	0.85	1.20
Other charges			
VAT	0.02	0.14	0.19
Total	0.3	1.99	2.79

Source: Tariff reference date: 1 Jan 2016, WAJ Amman, Jordan, 2016

Table 5.7 Zoning scheme for groundwater protection zones

Zone	Size	Land-use restrictions
Zone I	Springs: 50 m upstream, 15 m lateral, 10 m downstream Wells: 25 m upstream, 15 m lateral and downstream	Fenced off; access only for personnel of water works and management and control agency (government purchases property)
Zone II	50-day line; but: maximum of 2 km in upstream and 50–150 m in downstream directions in karst areas: areas classified as highly and very highly vulnerable (groundwater vulnerability map); but: maximum of 2 km in upstream and 50–150 m in downstream directions	Existing settlements and organic permitted; new development only with permission granted by licensing committee; existing land uses which generate waste and sewage water that may negatively affect the environment have to apply rules of best management practice (BMP), which will be issued by the government; construction of wastewater collection and treatment systems in this zone will be given priority
Zone III	Entire contribution zone	All types of land use are permitted if they are conducted in accordance with the rules of best management practice

Source: El-Naqa and Al-Shayeb (2009)

prevent water-quality-related health problems. Some costs are the implementation and operational costs of the groundwater-protection zones (Zones 1, 2, and 3) while the main costs are for rehabilitating the existing sewer system and/or connecting the unconnected houses in Zones 2 and 3 (Tables 5.6 and 5.7) (El-Habbab 2013).

5.6.1 Brine Disposal

The Red-Dead Sea project will be applied through three phases: (a) water transfer from the Red Sea to the Dead Sea, (b) hydropower generation and a reverse-osmosis desalination facility (capacity of 850 MCM/year), in addition to 30 MCM of desalinized water in other project, and (c) freshwater transmission system.

The quantity of brine and seawater from this project will be about 110–220 MCM/year. The water will be discharged to the Dead Sea through a pipeline of 56-km length (Rabadi 2015). On the other hand, there are considerable quantities of brackish groundwater in some basins, mainly in the central Jordan Valley and the Zarqa Aquifer in the Hisban and Kafrein areas.

5.7 Water–Energy Nexus

As a result of scarce surface-water resources, hydropower is limited through the King Talal Dam, which has a capacity of 5 MW, and the Aqaba thermal power station, which also has a capacity of 5 MW.

Electrical power generation in Jordan relies predominantly on fossil fuels with significant impact on the environment through harmful greenhouse gases (GHG) such as CO_2 and NO_x. Today, photovoltaic technology can produce solar power at a fixed rate of 0.05–0.08 JOD/kWh calculated for a 20-year project. This rate is lower than the cost of power produced with conventional methods. Photovoltaic power supply system represents an opportunity for the water sector to significantly decrease operational expenses and to mitigate the effects of energy price volatility, which largely depend on fluctuating fossil fuel prices. The water sector involves an energy-extensive operation by deploying large water pumping, boosting, treatment, and distribution facilities. The power requirements only for water pumping in 2014 amounted to about 15% of the total power production of Jordan with a total amount of 1592 GWh. The specific energy consumption for the same year was 7.51 kWh/m^3 (billed) for the Water Authority of Jordan (WAJ), mainly for municipal water supply and wastewater, and 0.274 kWh/m^3 (billed) for the Jordan Valley Authority (JVA), mainly for irrigation and industrial use in the Jordan Valley. The weighted average consumption for the public-sector water facilities is 4.31 kWh/m^3 (MWI 2016a, b).

The water sector, among all other power-consuming sectors in the country, will directly benefit from the implementation of the national energy strategy, which states that renewable energy shall contribute with a rate of 7% to the overall energy mix by 2015 and 10% by 2020. The introduction of renewable energy technologies into the water sector shall lead to the following results: (ibid)

- Supply of power at stable and low rates leading to reducing energy prices volatility
- Reduction of the sector dependence on fossil fuels
- Reduction of water-pumping costs
- Enabling long-term planning of water supply
- Reducing CO_2 emissions making it a high ecological value option, too.

5.8 Special Issues

Many water resources are transboundary and, as such, especially in semi-arid regions such as Jordan, are a basis for conflict. Wolf (1996) showed that most past attempts – from the early 1950s to 1991 – to resolve water issues in the Nile, Tigris-Euphrates, and Jordan Basins without considering socio-political implications failed. Political and water-scarcity issues cannot be addressed in isolation.

Many policymakers consider water shortages to be an intractable problem with multi-dimensional conflicts between countries on Jordan's borders. Many actions, such as water-management treaties, are taken to resolve the conflicts about using water from major rivers in the area.

Jordan has concluded two bilateral water agreements with Israel and Syria to manage shared water resources in the Jordan basin. Jordan has not benefited much from either. This is in part due to its weak strategic position against more powerful interlocutors and Jordan's little success in implementing many of the provisions of the agreements. The Peace Treaty of 1994 between Jordan and Israel calls for desalination projects on the Lower Jordan River but these have yet to be built. Further, diversion of 60 MCM from winter floodwaters of the Yarmouk River to Lake Tiberias for use by Jordan has not materialized either (Haddadin 2006). Jordan also claims that it has been able to access less than half of its share of flow from that river (Haddadin 2006). The Agreement of 1987 focuses on establishing the Al-Wehdah (Unity) Dam on the Yarmouk River. The latter has an annual gross flow of 110 MCM and a capacity to generate 18,800 kWh of power (MWI 2009a). However, because of excessive depletion of the Yarmouk's surface and groundwater, the water retained in the dam has been well below its 110 MCM capacity, sitting at little more than 18 MCM since its construction in 2006 (Namrouqa 2010). Even after the 1987 Agreement, the Syrians increased damming of the four recharge springs of the Yarmouk and have increased groundwater drilling in the river basin (Al-Kloub and Shemmeri 1996; Haddadin 2006; MWI 2007), leading to significant reductions in base flow along the Jordanian/Syrian border (Haddadin 2006; MWI 2007). Base flow is estimated to have dropped to 2 cubic meters per second in 2000, and to 0.9 cubic meter per second in 2008, compared to 5–7 cubic meters per second in the 1950s (Haddadin 2006; Ministry of Water and Irrigation, Jordan 2007; Namrouqa 2010).

The Jordan River Basin is one of the main sources of conflict in the Middle East region. It is shared by three countries: Jordan, Palestine, and Israel. One suggested solution for the conflict is to trade water according to market mechanism, but this procedure has serious problems because property rights are weakly defined and because of distrust between the parties leads to inefficient and environmentally damaging outcomes. To solve this problem, water-use rules are suggested to prevent over-extraction by some partners (Luterbacher and Wiegandt 2002).

5.8.1 Regional and International Water Issues

The Jordan River system and associated aquifers are important sources for Jordan and Israel. The Upper Jordan River is fed by three sources – the Dan, the Hasbani, and the Banias, which empty their water in the Sea of Galilee. Yarmoouk River, which has its source from Syria, is considered one of the boundaries between Jordan Israel and Syria; it empties its water in Jordan Valley near north (Adasyah Black 2010).

The majority of water resources in Jordan, such as the Disi aquifer and major rivers, are shared, originating outside the Kingdom. Transboundary water cooperation on the Jordan River Basin between Jordan and Israel scored 56.67 under the Water Cooperation Quotient (WCQ) 2017 that quantifies the quality of cooperation within transboundary river basins on a global basis, which means that the two countries have a relatively peaceful and stable relationship with each other in this respect. But in the case of the Yarmouk River Basin, the WCQ 2017 explained that since 2011, Syria has been unable to attend to its transboundary water relations due to the protracted armed conflict, indicating that any two countries not engaged in active water cooperation "do not necessarily go to war" (Namrouqa 2017a, b).

5.9 Other Issues

Social, Equity, and Institutional Performance One of the main, weak institutional performances in Jordan's water sector is the overlapping responsibilities for the Ministry of Water and Irrigation (MWI) as well as the government-owned Water Authority of Jordan (WAJ) and Jordan Valley Authority (JVA). The MWI has no responsibility for the water supply while the WAJ has both supply and retail roles which hinder water management in the country. The JVA is responsible for carrying out the socioeconomic development for Jordan's side of the Jordan Valley and is responsible for land distribution in the Jordan Valley. Another institutional problem for Jordan's water sector is the lack of agreement about using multiple data sources and over-staffed institutions (Yorke 2016).

In this respect, it is recommended to achieve the following activities (USAID 2017):

Strengthening and consolidating authority for water planning and management to address the over-extraction of groundwater improve sector planning and ensure better quality of data used in decision making. Solutions include capacity building, policy reform, process improvements, legal reforms, and institutional restructuring;

Reorganizing WAJ to focus on its core mandate (sector investment and bulk water source development/supply) to improve operational efficiency and better plan for future water supply and wastewater service needs;

Focusing on capacity to manage new national water supplies and removing institutional conflicts of interest between bulk water supply, utility oversight, and retail service delivery;

Strengthening utility management to support service improvement, further corporatization, and management/fiscal/operational independence; Improving water utility regulation to enhance the monitoring of utilities' financial and technical performance;

Strengthening water-user associations and the Jordan Valley Authority to further separate bulk and retail water management to improve irrigation water management and services across the Jordan Valley;

Conducting technical studies to provide analysis for better policy and water management decision making.

The National Water Strategy 2016–2025 was established to revise the institutional and legal frameworks in order to enhance the workable management activities for Jordan's water sector. Moreover, the strategy's main actions are: (a) including provisions for climate change; (b) focusing on water economics and financing; (c) ensuring the sustainability of overexploited groundwater resources; (d) adopting new technologies and available techniques, such as decentralized wastewater management; and (e) reusing treated wastewater (MWI 2016a, b).

The Millennium Challenge Corporation (MCC) was established in Jordan to increase the supply of fresh water for homes and businesses in the Zarqa Governorate, aiming to improve economic growth and the quality of life through three interrelated projects: (a) modernizing the water-supply infrastructure to improve the freshwater delivery efficiency in the city, (b) developing the sewerage in the project area to increase the wastewater-collection volume, and (c) enhancing the As-Samra wastewater treatment plant's capacity to provide treated wastewater to farms (Millennium Challenge Corporation 2010).

5.10 Conclusions

Jordan's water resources, both surface water and groundwater, mainly depend on rainfall. Groundwater from 12 groundwater basins accounts for about 54% of Jordan's water supply. Surface-water resources in Jordan have two principal parts: base flow and flood flow. Base flow is derived from groundwater drainage through springs. Surface water forms about 37% of the total water supply which is developed through 15 water basins that are distributed across the country.

Jordan is a water-poor country. Water scarcity threatens Jordan's development. Only 0.7% of the country has annual precipitation of more than 500 mm. The variable and low rainfall amounts with the high evaporation rate and droughts all contribute to low water-resource reliability and availability.

On the supply side, Jordan should improve water-supply and wastewater treatment infrastructure. Introduce affordable technologies for utilities, communities, and households to reduce water losses, private sector participation in infrastructure investments, reduce water losses, strengthen water sector institutions and policies, and encourage best commercial practices in water utilities. The Red Sea-Dead Sea (RSDS) Phase II project, which will be implemented in the near future, is conceived to address the challenges associated with the Dead Sea's declining water level and Jordan's ongoing water crisis.

On the demand side, MWI is encouraging the installation of Water Saving Devices (WSD) through the introduction and enforcement of a revised building code for new buildings. Retrofitting of existing buildings has been piloted in two areas through sponsorships.

A great deal of research covering different aspects of water demand was conducted for Jordan's water sector. On the supply side, because Jordan is characterized by limited water resources, Jordan has to emphasize water-resource management to meet the increased demand.

Jordan's current water policy requires a strong redirection toward water demand management. Actual implementation of the plans in the national water strategy (against existing oppositions) would be a first step. However, more attention should be paid to reducing water demand by changing the consumption pattern of Jordanian consumers. Moreover, unsustainable exploitation of the fossil Disi aquifer should soon be halted and planned desalination projects require careful consideration regarding the sustainability of their energy supply. Moreover, water policy encourages water harvesting, conserving, and protecting resources, while the water substitution and reuse policy proposes the reuse of treated wastewater in irrigation in order to enable the freeing of fresh water for municipal uses.

In irrigation it is recommended to consider water-saving technologies, replacement of groundwater with treated wastewater for farms located not far away from the existing or planned wastewater plants.

References

Abu-Awwad, A., & Blair, S. (2013). Economic efficiency of water use by irrigated crops in Al'Azraq area. *Jordan Journal of Agricultural Sciences, 9*(4), 525–543.

Al-Ansari, N., Alibrahiem, M., Alsaman, M., & Knutsson, S. (2014). Water demand management in Jordan. *Engineering, 6*, 19–26.

Albakkar, Y. (2014). *An integrated approach to wastewater management and reuse in Jordan – A case study of Jordan Valley*. A thesis submitted to the Committee of Graduate Studies in partial fulfillment of the requirements of the degree of masters of arts in the Faculty of Arts and Science, Trent University, Canada.

Al-Karablieh, E., & Salman, A. (2016). *Water resources, use and management in Jordan – A focus on groundwater* (IWMI Project No 11).

Al-Kloub, B., & Al-Shemmeri, T. (1996). Application of multi-criteria decision aid to rank the Jordan-Yarmouk basin co-riparians according to the Helsinki and ILC Rules. In J. A. Allen

& J. H. Court (Eds.), *Water, peace, and the Middle East: Negotiating resources in the Jordan Basin*. London/New York: I.B. Tauris.

Al-Mutaz, I. (2005). *Hybrid RO MSF desalination: Present status and future perspectives*. International Forum on Water–Resources, Technologies and Management in the Arab World, 8–10 May 2005, University of Sharjah, Sharjah, United Arab Emirates.

Al-Zu'bi, Y. (2007). Application of multicriteria analysis for ranking and evaluation of waste water treatment plants and its impact on the environment and public health: Case study from Jordan. *Journal of Applied Sciences Research, 3*(2), 155–160.

Ammary, B. (2007). Wastewater reuse in Jordan: Present status and future plans. *Desalination, 211*(1–3), 164–176.

Arabiyat, S. (2005). *Water price policies and incentives to reduce irrigation water demand: Jordan case study*. CIHEAM, Italy. http://om.ciheam.org/article.php?IDPDF=5002254. Accessed 16 Sept 2017.

Black, E. (2010). Water and society in Jordan and Israel today: An introductory overview. *Philosophical Transactions of the Royal Society A, 368*, 5111–5116. https://doi.org/10.1098/rsta.2010.0217.

Blogger, G. (2012). *Water shortages in Jordan*. State of the Planet, Earth Institute, Colombia University.

Department of Statistics (DOS). (2017). *Demographic statistics*. Amman: Department of Statistics.

Durham, B., Etienne, S., Gaid, A., & Luck, F. (2005). Le recyclage de l'eau et la gestion integree du petit cycle de l'eau. 84 Congres de l'ASTEE, May 2005, at Paris, France.

El-Habbab, H. (2013). *Socio-economic and environmental evaluation of Wadi Shuieb decentralized wastewater treatment (DWWT) plant and Hazzir groundwater protection zone in Jordan*. MSc. theses, University of Jordan, Amman, Jordan.

El-Naqa, A., & Al-Shayeb, A. (2009). Groundwater protection and management strategy in Jordan. *Water Resources Management, 23*, 2379–2394.

El-Sadek, A. (2010). Water desalination: An imperative measure for water security in Egypt. *Desalination, 250*(3), 876–884.

Faruqui, N. (2002). *A brief on wastewater treatment and reuse for food and water scarcity*. http://www.idrc.ca/en/ev-44039-201-1-DO_TOPIC.html. Accessed 22 June 2017.

FAOSTAT. (2017). *Data, crops, food and agricultural Organization for United Nation*. Rome, Italy.

Hadadin, N., & Tarawneh, Z. (2007). Environmental issues in Jordan, solutions and recommendations. *American Journal of Environmental Sciences, 3*(1), 30–36.

Haddadin, M. (2006). Compliance with and violations of the unified/Johnston Plan of the Jordan Valley. In D. Hambright, J. Rageb, & J. Ginat (Eds.), *Water in the Middle East: Cooperation and technological solutions in the Jordan Valley*. University of Oklahoma Press.

Hjazi, J. (2010). *Water demand management in Jordan-case study*. Presented for the fifth partner forum on water governance in the MENA Region. Ministry of Water and Irrigation- Jordan.

Hoekstra, A. Y., & Mekonnen, M. M. (2012). The water footprint of humanity. *Proceedings of the National Academy of Sciences, 109*, 3232–3237.

Hussein, I. A., Abu Sharar, T. M., & Battikhi, A. M. (2005). Water resources planning and development in Jordan. *Food security under water scarcity in the Middle East: Problems and solutions:* 183–197. http://om.cih eam.org/article.php?IDPDF=5002212

International Resources Group (IRG). (2013). *National strategic wastewater master plan*. USAID.

Klinger, J., Grimmeisen, F., & Goldscheideustainable, N. (Eds.). (2012). *Management of available water resources with innovative technologies*. Karlsruhe Institute of Technology (KIT), Institute of Applied Geosciences, Division of Hydrogeology.

Luterbacher, U., & Wiegandt, E. (2002). Water control and property rights: An analysis of the Middle Eastern situation. In M. Beniston (Ed.), *Climatic change: Implications for the hydrological cycle and for water management. Advances in global change research* (Vol. 10). Dordrecht: Springer.

Ministry of Water and Irrigation. (2007). *Germany will carry out a study for Yarmouk River to allocate water rights to Jordan and Syria.* http://www.mwi.gov.jo/mwi/new_Germany.aspex. Accessed 4 June 2017.

Ministry of Water and Irrigation. (2009a). *Water for life Jordan's water strategy 2008–2022.* Amman: The Hashemite Kingdom of Jordan.

Ministry of Water and Irrigation. (2009b). *Jordan.* http://www.mwi.gov.jo/English/MWI/Pages/Projects.aspx. Accessed 4 June 2017.

Ministry of Water and Irrigation. (2015a). *Annual water report 2015.* The Hashemite Kingdom of Jordan, Amman, Jordan.

Ministry of Water and Irrigation. (2015b). *Jordan water sector facts and figures.* Amman, Jordan: The Hashemite Kingdom of Jordan.

Ministry of Water and Irrigation. (2016a). *Energy efficiency and renewable energy policy.*

Ministry of Water and Irrigation. (2016b). *Water demand management in the context of water services Jordan.*

Ministry of Water and Irrigation. (2016c). *Water sector capital investment plan, 2016–2025.*

Ministry of Water and Irrigation (MWI). (2016a). *National water strategy of Jordan, 2016–2025.* Amman: The Hashemite Kingdom of Jordan.

Ministry of Water and Irrigation (MWI). (2016b). *Water re-allocation policy.*

Ministry of Water and Irrigation (MWI). (2017). *Annual Report.* Amman, Jordan.

Millennium Challenge Corporation. (2010). *Jordan compact.* https://www.mcc.gov/where-we-work/program/jordan-compact.

Myszograj, S., & Qteishat, O. (2011). *Operate of As-Samra Wastewater Treatment Plant in Jordan and Suitability for Water Reuse.* Inżynieria i Ochrona Środowiska 2011, t. 14, nr 1, s. 29–40.

Namrouqa, H. (2010). Cheap water for agriculture exacerbating shortage. *The Jordan Times.* http://www.jordantimes.com/index.php?news=23167 Accessed 4 Oct 2017.

Namrouqa, H. (2017a). Jordan can reduce agricultural water use by a third, research finds. *Jordan Times.*

Namrouqa, H. (2017b). Water cooperation on Jordan River Basin between Jordan, Israel scores 56.67. *Jordan Times.*

Odom, O., & Wolf, A. (2011). Institutional resilience and climate variability in international water treaties: the Jordan River Basin as "proof-of-concept". *Hydrological Sciences Journal, 56*(4), 703–710. Water Crisis: From Conflict to Cooperation.

Ornwipa T. and Rodriguez, C. (2008). Desalination in the red-Dead Sea conveyor project. http://courses.washington.edu/cejordan/Desalination%20in%20the%20RDSC%20Project.pdf.

Permaculture Research Institute. (2005). *Use of permaculture under salinity and drought conditions.* https://permaculturenews.org/2005/02/01/use-of-permaculture-under-salinity-and-drought-conditions/. Accessed 28 Aug 2017.

Qtaishat, T. (2012). Water-supply augmentation options in water-scarce countries. *International Journal of Water Resources and Environmental Engineering, 4*(8), 263–269.

Qtaishat, T. (2013). Impact of water reallocation on the economy in the Fertile Crescent. *Water Resources Management.* https://doi.org/10.1007/s11269-013-0379-z.

Qtaishat, T., Al-Karablieha, E., Salman, A., Tabieh, M., Al-Qudah, H., & Seder, N. (2016). Economic analysis of brackish-water desalination used for irrigation in the Jordan Valley. *Desalination and Water Treatment.* www.deswater.com. https://doi.org/10.5004/dwt.2017.20435.

Rabadi, A. (2015). The Red Sea–Dead Sea desalination project at Aqaba. *Journal Desalination and Water Treatment, 57*(2), 48–49.

Saidam, M., Epp, C., & Papapetrou, M. (2009). Appraisal of institutional and policy framework conditions for the use of autonomous desalination units in Jordan. *Desalination and Water Treatment, 5*(2009), 111–118.

Seder, N., & Abdel-Jabbar, S. (2011). *Safe use of treated wastewater in agriculture- Jordan case study.* Prepared for ACWUA, Amman, Jordan.

Shatanawi, M., Fardous, A., Mazahrih, N., & Duqqah, M. (2005). *Irrigation systems performance in Jordan.* Mediterranean Agronomic. Institute of Bari, Italy.

Smith, M., Cross, K., Paden, M., & Laban, P. (2016). *Spring – Managing groundwater sustainably*. Gland: IUCN.

Trojan, F., & Morais, D. C. (2012). Using ELECTRE TRI to support maintenance of water distribution networks. *Pesquisa Operacional, 32*(2), 423–442. https://doi.org/10.1590/S0101-74382012005000013.

UNESCO. (2012). *Managing water under uncertainty and risk: The United Nations world water development report 4*. Paris.

United Nations. (2010). *United Nations salaries: Allowance benefits and job classification (salary scales for staff in general service and related categories)*. http://www.un.org/Depts/OHRM/salaries_allowances/salaries/. Accessed 22 June 2017.

USAID. (2012). *A review of water policies in Jordan and recommendations for strategic priorities*. http://pdf.usaid.gov/pdf_docs/PBAAE636.pdf. Accessed 16 Aug 2017.

USAID. (2013). *Jordan Valley water users associations (WUAs): Future roles and responsibilities report*. http://pdf.usaid.gov/pdf_docs/PA00JRPT.pdf. Accessed 16 Aug 2017.

USAID. (2017). *Institutional support and strengthening program (ISSP) in Jordan*.

Van Den Berg, C., Al Nimer, A., Fileccia, S., Gonzalez, T., Maria, L., & Wahseh, S. (2016). *The cost of irrigation water in the Jordan Valley*. Water partnership program (WPP). Washington, DC: World Bank Group. http://documents.worldbank.org/curated/en/275541467993509610/The-cost-of-irrigation-water-in-the-Jordan-Valley. Accessed 25 July 2017.

Wolf, A. (1996). *Middle East water conflicts and directions for conflict resolution* (Food, Agriculture, and the Environment Discussion Paper 12). International Food Policy Research Institute, Washington, DC.

World Bank. (2007). *Making the most of scarcity: Accountability for better water management in the Middle East and North Africa. MENA development report*. Washington, DC: World Bank.

World Bank. (2016). *Worldwide governance indicators 2016*. Washington, DC. http://info.worldbank.org/governance/wgi/index.asp. Accessed 28 July 2017.

Yorke, V. (2016). Jordan's shadow state and water management: Prospects for water security will depend on politics and regional cooperation. In R. Hüttl, O. Bens, C. Bismuth, & S. Hoechstetter (Eds.), *Society – Water – Technology. Water Resources Development and Management*. Cham: Springer.

Dr. Tala Qtaishat earned her PhD in Natural Resources Management with major on Water Resources Economics and Policy in 2011 from North Dakota State University, USA. She is Assistant Professor at the Department of Agricultural Economics and Agribusiness Management, College of Agriculture, University of Jordan. Her main research interest is in the water planning and policy fields. She is currently carrying on a research project on water use efficiency for improving water productivity in the Jordan Valley. The project is funded by the European Commission. She has published five papers on water economics/policy and presented several papers in international conferences. She is Member of the International, American, and Canadian Water Resources Associations.

Chapter 6
Oman Water Policy

Slim Zekri

Abstract Oman is a country under severe water stress. Currently Oman produces around 1 Mm^3/day of desalinated seawater for urban purposes to expand supply. This policy was partially imposed by the irregularity of rain and the concentration of the population on the coastal areas. Most of the conventional water resources are in the form of groundwater and are used in the agricultural sector. Abstractions from wells are subject to licenses. But licenses so far do not carry any limits. The result is a race for water with overabstractions in the coastal areas causing seawater intrusion and damage to the aquifers. The government is planning to introduce progressively water quotas to farmers and monitoring through smart meters and online system. Large volumes of tertiary treated wastewater are produced daily and are only partially reused for landscaping. There is a mismatch between the willingness of farmers to pay for treated wastewater and the price set by the public authority leading to a limited demand. The actual context of free and unlimited access to groundwater does not encourage to shift the demand toward high-quality treated wastewater. Plans are being considered for recharging some of the aquifers with the treated wastewater. Irrigation efficiency improvements have been observed mainly for vegetable producers where the adoption of irrigation technology resulted in higher revenues and lower labor costs. Urban water prices are at 1/3 of their costs discouraging water saving and adoption of water saving/recycling devices at homes or industries. Urban water security is being addressed by aquifer storage and recovery techniques using excess winter desalinated water.

Keywords Groundwater · Desalination · Climate change · Pricing

---

S. Zekri (✉)
CAMS, Department of Natural Resource Economics, Sultan Qaboos University, Al-Khod, Sultanate of Oman
e-mail: slim@squ.edu.om

© Springer Nature Switzerland AG 2020
S. Zekri (ed.), *Water Policies in MENA Countries*, Global Issues in Water Policy 23,
https://doi.org/10.1007/978-3-030-29274-4_6

6.1 Introduction

The rapid population increase and the extremely harsh climate conditions characterized by an erratic precipitation, high temperatures, and high evaporation rates are among the major factors that affect the water sector in Oman. The population reached 4.4 million in 2016 with expatriates representing 45% of the total population. On average, the population grew at 5.6% in 2016. The growth rate of expatriates reached 9.1% the same year. The local population growth rate is still on the high side with 3.5% (NCSI 2017). Furthermore, rural migration, concentration of the population in the coastal cities, and the economic growth exacerbate the pressure on urban water systems.

The annual average precipitation in the coastal areas of Oman is around 100 mm/ year. Evapotranspiration is estimated at 2000 mm a year, exceeding average rainfall by 20 times. In the mountain areas, where most of the recharge of aquifers comes from, rainfall reaches 250 mm/year. The mountains are rocky with fractures allowing rainfall to penetrate in the aquifers while the surface flows and run-off recharge the coastal aquifers directly through the Wadis' beds and through recharge dams. The conventional water resources represent 87% of the nation's water resources whereas nonconventional water resources including desalinated water and treated wastewater represent nearly 13%. Groundwater represents 94% of the conventional water sources (MRMWR 2013). Given the rainfall irregularity and the mismatch of the locations of aquifers and cities, most of the urban water is thus supplied by desalination plants. This has been facilitated so far by the availability of gas and oil resources for desalination purposes. While the urban water falls under the responsibility of the Public Authority for Electricity and Water, the agricultural water falls under the responsibility of the Ministry of Regional Municipalities and Water Resources (MRMWR) and the Ministry of Agriculture and Fisheries Wealth (MAF). Agriculture is the highest user of the conventional water in the country to produce mainly dates, forage crops, and vegetables. The country depends heavily on imports for food and will remain so in the future given the small annually renewed water resource, estimated at an average 1300 Mm^3/year. Water policy in Oman is predominantly under the command and control approach. Historically, however, water governance was initially decentralized and community based with Aflaj as a millenarian successful example of community water management. The current trends of water policy show a trend of swinging back to the allocation of water rights, learning from the Aflaj long lived experience.

6.2 Administrative, Legal, and Institutional Aspects of the Water Sector

Several administrations are involved in the regulation and management of water resources in Oman. These entities have evolved over time and the current picture is as follows. For the conventional water resources, the two main bodies are the

6 Oman Water Policy

MRMWR and the MAF. The MRMWR is in charge of: development of policies plans and programs necessary for the water sector; mMonitoring, evaluation, and development of water resources to achieve a balance between water uses and renewable resources; conservation of water resources from depletion and the rationalization of water consumption in coordination with the concerned authorities.

The MAF is responsible for the use of water for irrigation purposes only. MAF provides farmers with extension services, advice on crop mix, and subsidies to adopt modern irrigation systems at farm level. The MAF plans for innovation in the traditional Aflaj irrigation systems and undertakes pilot studies for reforms. Irrigation water user associations are numerous and quite spread in the whole country. These associations are called Aflaj. They have their own water regulations and management rules. The Ministry of Environment and Climate Affairs is the third ministry with responsibilities on water. It has the responsibilities of developing and delivering policies and plans for protecting the environment, controlling pollution and nature conservation, and climatic affairs. The majority of its activities focus on developing legal and regulatory standards and permitting and controlling procedures for protecting the environment from different activities, including discharges into and abstractions from water and air emissions from water production processes. Both the Ministry of Environment and Climate Affairs and the MRMWR are responsible for protecting the water resources from pollution, and it is not always clear where the role of one ministry ends and the other begins.

Most of the water for urban uses is nonconventional water. The urban fresh water sector is governed mainly by the Public Authority of Electricity and Water (PAEW) and the Oman Power and Water Procurement Company (OPWP). These two public agencies are responsible for the planning, finance, procurement, and supply of desalinated water in the whole country. Both public and private desalination plants are under the umbrella of the PAEW. The PAEW is responsible for the pricing of urban water. The Oman Power and Water Procurement Company (OPWP) is a government-owned holding company under the umbrella of the Ministry of Finance. The OPWP is in charge of ensuring supply of bulk desalinated water to the PAEW. The OPWP is in command of planning and contracting the private sector companies producing desalinated water.

The treated wastewater is taken care of by two semipublic companies. Oman Wastewater Company (Haya) is the Sultanate's largest state-owned company in charge of building and operating the wastewater network and sewage treatment plants in ten Governorates of the country. The Salalah Wastewater Company has the same role as Haya but limited to the southern part of the country and in the Governorate of Salalah only. Both companies are in charge of extending the collection network, treatment, and distribution of the treated wastewater.

Oman Water Society (OWS) is a nonprofitable nongovernmental organization. It gathers professionals with interest in the water sector as a whole with an integrated perspective. OWS organizes workshops and seminars to raise awareness about the challenges facing the water sector and is a platform for the exchange of ideas and technology advances among experts and the public in general.

The water sector regulations in Oman are in the form of Royal Decrees and Ministerial Decisions. A number of Royal Decrees and Ministerial Decisions were issued to protect, conserve, and improve the management of the water resources. The most salient laws are reported below. The Royal Decree 82/1988 is considered the most important piece of legislation on water resources. The Royal Decree 82/1988 considered for the first time water as a national resource. It is a sort of nationalization of the resource giving full right and power to the government to make decisions and choices on the water sector. It states that "...the water of the Sultanate of Oman is a national resource to be used according to the restrictions made by the government and organizing its optimum utilization in the interest of the state based on comprehensive development plans." In 1989, the government issued the Royal Decree 72/89 encouraging the adoption of modern irrigation systems to rationalize water use, stressing the need to increase crop yields, and enhancing the quality of the products. The decree allowed the government to provide a fund to subsidize the cost of modern irrigation systems as an incentive for farmers to speed up the adoption process. The same decree is still in use for the incentives.

Given that 94% of the natural water resources are in aquifers, the Royal Decree 29/2000 is a Water Resources Protection Act that emphasizes the regulations of wells and Aflaj, permits, and maintenance and regulates desalination units on wells. The Royal Decree 114/2001 on environment conservation and prevention of pollution regulates the disposal of solid and hazardous waste, pollution control, and the release of permits for dumping untreated wastewater. Royal Decree 115/2001 refers to organizing the disposal of liquid and solid waste products. In 2001, a series of Ministerial Decisions established water supply well field protection zones in several regions of the Sultanate.

6.3 Agricultural Water

All agricultural activities depend totally on irrigation, except the range lands that partially feed some of the animals. Currently 83% of the conventional water resources are used for agricultural purposes (MRMWR 2013). Most of the water is used to produce dates and fodder crops. Seventy nine percent of the cropped area is allocated to four crops: date palms, Rhodes grass, alfalfa, and banana (see Fig. 6.1). Not only these crops have the highest water requirement per hectare, they have also the lowest net return per cubic meter (FAO 2008). The reasons why farmers are producing these crops are that water is free and farmers pay the cost of pumping only. The only limitation for pumping is the well's yield. As for forages, the crops are easy to produce with automatic irrigation, are risk free in terms of yield variability, and market is secured both locally and export for neighbor country. The date palms constitute a traditional crop and are a low input activity. On the opposite, high value crops such as vegetables, which require far less water for their production, are risky in terms of both yield and price variability and are input intensive. Joseph (2017) estimated the virtual water export from dates and alfalfa to be 40 and

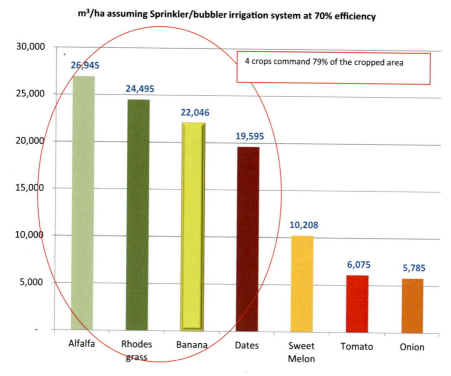

Fig. 6.1 Water consumption in m³/ha assuming Sprinkler/bubbler irrigation system at 70% efficiency. (Source: Self elaboration)

19 Mm³/year, respectively. The net sale for dates was 0.107 OR/m³ for the whole chain and for alfalfa it was 0.185 OR/m³. Changes to farm management and export policies are required to contribute to the reduction of the water deficit.

The agricultural water is supplied through two main systems in which traditional Aflaj account for 32% of total groundwater while the remaining amount is pumped from private wells (MRMWR 2008). The total demand for agricultural water has increased annually at a rate of 3.3% during the period 2000–2011. As a consequence, groundwater deficit increased from 285 million m³ in 1990 to 316 million m³ in 2011 with alarming levels of seawater intrusion in the coastal aquifers (MRMWR 2013). Master water plans since 2005 recommended the management of groundwater resources to control saline water intrusion as an absolute priority for the country. Agricultural water demand is not evenly distributed. The largest volumes of agriculture water are abstracted from the Batinah coastal aquifers with an estimated demand of 730 million m³ in 2011 representing 36% of the total demand in the country. The agricultural area in the Batinah region is the largest and estimated at 73,626 Feddan (acres) covering 53% of the total cropped area in the Sultanate (MAF 2013). Al Dakhlia region is the second highest region in water demand with an estimated volume of 267 million m³.

Table 6.1 Oman's water administration

Organization	Subsector	Responsibilities
Ministry of Regional Municipalities and Water Resource (MRMWR)	Water resources	Regulation
		Water resources management
		Water resources assessment
		Water resources development
Ministry of Agriculture and Fisheries (MAF)	Irrigation water	Irrigation water management
The Ministry of Environment and Climate Affairs	Water quality and discharge	Legal and regulatory standards
Public Authority of Electricity and Water (PAEW) (under Ministry of Finance)	Urban fresh water	Regulator for the urban water sector
		Direct water service provider
		Urban water supply and management to users
Oman Power and Water Procurement Company (OPWP) (under Ministry of Finance)	Desalinated water	Ensures bulk supply of desalinated water to the PAEW
		Planning and contracting new capacity from the private sector companies
Falaj organizations (Water users Associations)	Community	Falaj water allocation
		Falaj protection
		Falaj maintenance
		Falaj water distribution
Oman Wastewater Company (Haya)	Wastewater	Collection, treatment, and reuse of treated water
Salalah Wastewater Company	Wastewater	Collection, treatment, and reuse of treated water
Oman Water Society	Community water use	Water conservation

The total balance and distribution of water among Oman's regions are shown in Table 6.1, which shows the available conventional water resources supply and the total demand. The average available amount of groundwater in Oman is estimated at 1465 million m^3 annually, while the total demand is estimated at 1781 million m^3. The net water balance is consequently a deficit of 316 million m^3 that comes in the form of seawater intrusion. In general, 50% of Oman's regions have a positive water balance. The other six governorates have a negative balance. The last row shows that less than 10% of the rainwater is captured into aquifers, despite the government's efforts of building recharge dams (Table 6.2).

The excessive abstraction of groundwater causes quality degradation, financial losses, and threatens sustainability of the aquifer usage in the future. Groundwater deficit or aquifer depletion has been triggered by the land distribution program, the introduction of diesel pumps, and provision of rural areas with electricity since the late 1980s.

The Oman National Water Resources Master Plan in 1991 (NWRMP) called for the reduction of 215 million m^3 in the abstraction of groundwater. The NWRMP

6 Oman Water Policy

Table 6.2 Water balance per governorate in million m³

Region	The volume of rainfall	Groundwater recharge	Groundwater uses	Excess/deficit
Musandum	333	57	31	26
Al Batinah North	1243	280	535	−256
Al Batinah South	329	104	195	−91
Muscat	518	56	80	−24
Al Burymi	569	92	67	25
Al Dhahirah	1423	108	250	−142
Al Dakhliyah	4627	219	267	−57
Sharqiah North	1136	122	177	−55
Sharqiah south	296	75	27	48
Masira	43	10.9	5.7	5.2
Al Wusta	1894	97.1	2.5	93.6
Dhofar	2390	244	133	111
Total	15,841	1465	1781	−316

Source: The Ministry of Regional Municipalities and Water Resources (2013)

recommended the adoption of modern irrigation systems, cropping of winter vegetables instead of date palm trees, implementation of suitable tariff system for all other purposes excluding the agricultural sector, and the reuse of treated wastewater for municipal irrigation and landscaping. The recommendations have been only partially implemented. For example, the modern irrigation system is currently covering around only 39% of all cropped areas given the limited funds for subsidies. Most landscape irrigation is undertaken using treated wastewater. Date palm and forage crops, requiring high volumes of water, are still the dominant crops. Finally, the most important recommendation regarding the monitoring of groundwater abstraction has not yet been implemented due to technical problems and reluctance to face the farmers with abstraction quotas. The MRMWR preference was to act on the supply side with the construction of 43 recharge and flood protection dams. Most of these dams are located along the Batinah coastal area and were designed to control floods and reduce seawater intrusion. The average captured volume that recharged the aquifers, during the period 1990–2010, was 50 million m³.

6.3.1 Aquifer Depletion

Farmers are extracting much more water than is renewed each year. As a consequence, some 144 Mm³ of seawater are flowing into Oman's aquifers each year. Up to now, 23,000 feddan of farmland have become too salty and have been abandoned (MAF 2012). If the current groundwater overabstraction continues, the aquifers will become increasingly salinized and the groundwater will be impaired.

One important step toward groundwater monitoring was undertaken in 1992 through the national well inventory project. The inventory allowed locating all the wells with their respective GPS coordinates, the installed pumps' type and horse power and abstraction purpose. The data base was used to estimate the total groundwater abstraction. The second step was the issuing of the Ministerial Decree that for the first time introduced the possibility of groundwater monitoring through individual quotas. The MD 264/2000 article (21.A) states that "The Ministry shall determine the quantity of water to be taken from each well". Even though the well inventory did not lead to measures to control abstraction, it had the merit to bring under strict monitoring the wells' deepening, the creation of new wells, or the replacement of existing wells. No new wells are allowed for agricultural uses unless to protect and preserve existing plantations or as a support for dried up Aflaj. However, new wells are allowed for municipal uses, petroleum, industrial, and touristic uses. This is an indicator that the new wells and water are allocated implicitly to high value uses. The sustained demand on new licenses is an indicator of the increased scarcity of groundwater as well as a proof of the negative impact of drought periods on the users. The licenses issued reached a peak during 2004 with 7129 licenses issued. This peak coincided with the low runoff in 2003 and 2004 (see Fig. 6.2). Thus, the demand for licenses depends on the drought periods and their duration. Every drought period is accompanied by new demands for wells' licenses.

The increase of abstraction affected the traditional Aflaj irrigation systems. Since 1990, the MRMWR counted some 1000 dried-up Falaj, out of a total of 4000. To mitigate the degradation of Aflaj the MRMWR supported 669 projects allowing farmer communities to dig new wells in order to maintain life in the rural areas. A study of a sample of 33 dried-up Falaj shows that the resulting impact was the loss of agriculture income, deterioration of lifestyle, and changes in land value. Around 16% of the families living around Falaj were obliged to relocate as result of the dry-up. The total losses at the national level related to Aflaj dry-up were estimated to 59 million Omani Rials per year (Zekri et al. 2012). The lack of implementation of

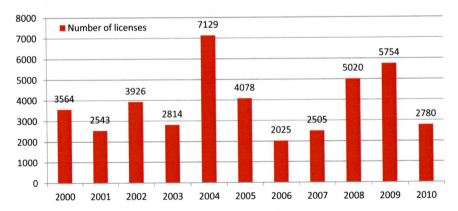

Fig. 6.2 Wells licenses issued by the MRMWR during the period 2000–2010 all over the Sultanate. (Source: The Ministry of Regional Municipalities and Water Resource 2013)

the law protecting the Aflaj (3.5 km exclusive protection zone around the mother well), which prohibits digging wells in the vicinity of Aflaj, is one of the main causes for the drawdown of the water table and hence the dry up of Aflaj.

Zekri et al. (2017) tested smart water meters on a pilot study that included 40 farmers and estimated the potential benefits of monitoring groundwater. The advantage of the smart water meters is that they have no moving parts and are monitored online. Optimization of abstraction over the next 70 years showed that a proper management of the groundwater requires a reduction of abstraction by 20% and changes in the crop mix that will improve the salinity of the aquifer in the long run compared to the current status and farmers' maximize profit. The recent sustainable agricultural and rural development strategy 2040 proposed to start the introduction of water metering in a large-scale area of 8000 farmers for a bigger test before generalization (FAO-MAF 2016).

6.3.2 Water Rights

Even though the Royal Decree 82/1988 considered water as a national resource, it excluded the Aflaj systems, which are still governed by the traditional water laws and rights. In Aflaj, users have access to irrigation water only if they own/rent a water right, while all community members have open access to Falaj water for domestic purposes. The traditional law protects the water rights for production purposes and protects the basic human needs for water. In small villages, domestic demand is often relatively negligible compared to agricultural water demand. The community members are allowed water for domestic purposes from the main channels but without connecting the houses by a network of pipes to avoid excessive use of water. This is a way of making access to domestic water costly to encourage water saving. The traditional laws reflect a vision of water rights based on both equitability and efficiency. Furthermore, public baths built on the Falaj sides are available for free for all users. The graywater flows from these baths are returned to the irrigation channel. This shows an advanced water management system totally adapted to the dry conditions of the country devised several hundred years ago.

The water rights in Aflaj are private, common, or quasi-public. Most of the water rights are private (88%), licensed, and can be traded. The common water rights are owned by the Falaj community. They are rented through lease auctions and the rent insures a sustainable management through the provision of continuous monetary flows for maintenance and operation. The last category is the quasi-public water rights that are owned by charity entities. This category of water rights is leased in short-period markets and the revenue is used to support social organizations. The common and quasi-public water rights represent 6% each (Zekri and Al-Marshudi 2008).

Water rights in Aflaj are millenarian and are characterized by separation from land, privacy, proportionality, seniority, and transferability. These characteristics fluidize the exchange of the rights freely among farmers. Water rights privacy has similar legal sides to any other private asset and it is recognized at the national level. The water rights are in the form of timeshare of the resource flow reflecting the principle of sharing existing supplies and shortages according to the water flow rate, allowing handling the uncertainty of water flow. The structural design of the Falaj's main channels grants the priority to use water for domestic purposes. The total dependence of domestic users on Falaj flow boosts the volunteering of community members to restore the water flow if interrupted in case of collapse of the channels. This contributes to the sustainability of the system by speeding up the repair and cuts labor costs in cases of big repairs since it is a win-win option. Users get the domestic water for free but they need to give hand to restore the system whenever needed.

The law maintained the existing seniority right through an exclusive zone surrounding the mother well to protect the Falaj ownership, since the aquifer source of groundwater is the same for both the Falaj and the wells. The system's sustainability depends extremely on the implementation of the seniority principle and groundwater abstraction regulation.

6.3.3 Reuse of Treated Wastewater

Haya Water is a semiprivate wastewater treatment company established in 2002. In 2014, the Omani Government assigned to Haya Water the responsibility of the development, execution, and management of wastewater facilities in the country, except the Governorate of Dhofar. Haya Water currently operates 57 sewage treatment plants at 44 different locations. Previously, Haya was responsible only for wastewater in Muscat and the MRMWR was in charge of the plants outside Muscat. Currently, the plants operated by Haya produce around 550,000 m^3/day of treated wastewater, and this volume is expected to increase to 660,000 m^3/day by 2020 as the number of connected houses to sewage network increases (Haya 2018).

In Muscat city, the amount of tertiary treated wastewater produced in 2017 reached 60 Mm^3, out of which 31 Mm^3 were recycled. Muscat Municipality buys 68% of the volume at a price of 0.100 OR/m^3 and uses it to irrigate public landscapes. So far citizens do not pay any municipal tax and Muscat Municipality is funded from the central government budget. This hinders the sustainable use of treated wastewater in the future given that the governmental funds depend heavily on nonrenewable oil income. The remaining uses are for industrial purposes or golf courses with varying prices. The agricultural uses of treated wastewater are negligible given the high price asked by Haya, far higher than the farmers' willingness to pay. The result is that 30 Mm^3 of high-quality tertiary treated wastewater end up in the sea while the country is under absolute water scarcity and a huge water deficit.

Zekri et al. (2016) estimated at 0.111 OR/m^3 the average farmers' willingness to pay for treated wastewater for irrigation. The study included both hobby and commercial farmers. Commercial farmers are willing to pay less than the average with 0.087 OR/m^3. Haya Water Company is demanding a price of 0.220 OR/m^3, which most of the farmers cannot afford as it is higher than their marginal benefit. Besides, treated wastewater is currently far more expensive than the groundwater open access resource, the cost of abstraction of which ranges between 0.005 and 0.023 OR/m^3. Under such conditions, the treated wastewater demand for agricultural purposes will be restricted only to hobby farmers who have seen their groundwater badly salinized.

On the other hand, households are charged an average price of 0.154 OR/m^3 far below the cost of treatment estimated at around 0.800 OR/m^3 for the city of Muscat. Haya owns the treated wastewater and sells part of it to cover partially the costs (19%). The difference between cost and revenue is covered by public subsidy estimated at 60% of the cost (Zekri et al. 2016).

Several policy changes are required to avoid the wasteful disposal of tertiary treated wastewater into the sea: (1) households and users should pay the full cost of the treatment, (2) groundwater quotas should be allocated to each user, (3) treated wastewater price should take into account the farmers' willingness to pay, and (4) considering artificial aquifers' recharge, using treated wastewater, similarly to the public investments in recharge dams. During the period 1990–2010 all the recharge dams allowed to capture variable quantities ranging from 122 Mm^3 in 2007 to 3.2 Mm^3 in 2008 with an average of 50 Mm^3/year. Aquifer storage and recovery of treated wastewater is independent from rainfall variability and produces reliable volumes annually (Al-Maktoumi et al. 2016). Recharge using treated wastewater might be less costly than the recharge via dams estimated at 0.300 OR/m^3, depending on the distance between the source and the aquifer. This will entail allocating certain aquifers to agricultural uses exclusively given the quality of injection water. El-Rawy et al. (2018) evaluated the impact of managed aquifer recharge using treated wastewater in a coastal aquifer. They found that Net Benefit Investment Ratio can reach up to 3.18 if the injection takes place within a limit of 1 km from the water source.

6.3.4 Pricing

Groundwater is mainly extracted through Aflaj and wells. Groundwater is a free access resource in Oman. Farmers based on individual wells pay only the cost of the wells, pumps, and electricity for abstraction. In Aflaj, water markets among farmers prevail. Marginal prices of water in Aflaj vary from year to year, between seasons, and among Aflaj. Zekri et al. (2006) reported marginal prices in the range of 0.005–0.025 OR/m^3.

Domestic water prices are in the form of two block tariff (0.440 OR/m^3 for consumption of less than 23 m^3/month and 0.550 OR/m^3 above). These prices have not changed since the year 1980, even though the second block was instituted in year 2000. Prices are uniform through all the country. Most urban water is desalinated seawater. According to the Public Authority for Electricity and Water (2016), the subsidy covers 2/3 of the water cost and is estimated at OR 186 million/year. Prices for commercial and governmental uses have been revised in 2016 and established at 0.780 OR/m^3 as one block tariff. The prices of water are still far below the average cost of production plus distribution estimated at 1.550 OR/m^3. The low prices to domestic and industrial users do not encourage users to save water or adopt water saving/recycling technologies. In fact, domestic water consumption in Oman is on the high side with 200 L/cap/day. Kotagama et al. (2016) found that the price elasticity of water for residents in villas is high and is −2.10 due to the outdoor uses. The authors highlighted the possibility of managing water demand in Muscat through modifying the price of water and reforming subsidies for domestic users. The protection of vulnerable sectors in the community is the main argument of keeping the domestic water and electricity prices at a low level (PAEW 2016). There are several international experiences where prices have been increased without hurting the low income households. Even the Omani experience of fuel price increases is a successful one that could be mimicked for the water sector. Since 2018, users are paying the full cost of fuel with a protection of the low income households, earning less than 600 OR/month. The subsidy is allocated using an electronic card to access fuel up to 200 L/month at a subsidized price. The fuel subsidy is now targeted exclusively to the low income households. Furthermore, Oman has also started a price policy reform in the electricity sector for large users. The new price scheme is Cost-Reflective Tariff intended to reduce the high-cost peak generation and additional investment for network. On the same line, there has been an increase in the price of water sold to both governmental and commercial sectors in 2016 as an initial step of performing the full cost recovery from both sectors. Introducing new pricing methods such as the uniform price with rebate or the more detailed method of the allocation-based pricing that takes into account the family size and the outdoor uses (Baerenklau et al. 2014) can help avoid the reform rejection by low income users.

6.3.5 Irrigation Efficiency

Wichelns (2014) affirms that "*In cases in which water is limiting, relative to land, the profit-maximizing solution will be the same as that which maximizes the average productivity of water.*" Surface or flood irrigation is the dominant irrigation system in Oman. The adoption of modern irrigation systems (sprinkler, drip, bubbler) increased from 6% in 1993 to 39.2% in 2013 as shown in Table 6.3. Despite the increased use of modern irrigation system, the abstraction from the aquifers has not been reduced. The main reason for this is the fact that groundwater is an open access resource and farmers pay only the cost of abstraction. The last column of Table 6.3 shows that the cropped area observed an increase of 16% during the period 2005–2013. In general, farmers who introduce modern irrigation systems tend to extend their cropped areas and/or further intensify the existing activities. The result is an improvement of water use efficiency without any water saving at the aquifer level. Farmers abstract water until the marginal cost of abstraction is equal to the

Table 6.3 Evolution of the area covered by MIS during the period 1992–2013 in Feddan

	Plantations	Vegetables	Field crops	Forage crops	Total crops under MIS	% cropped area using MIS	Total cropped area
Agr. Census 1992/1993	2440	1776	291	4326	8833	6.00	146,515
Agr. Census 2004/2005	5368	6384	1179	16,098	29,030	20.70	140,118
Agr. Census 2012/2013	11,913	22,286	2823	26,902	63,925	39.20	163,045
Rate of change 1992/2005	120%	259%	305%	272%	229%		−4%
Rate of change 2006/2013	122%	249%	139%	67%	120%		16%

Source: Agricultural Census 1992–93, MAF 2004, and 2012–2013

marginal benefit subject to the well daily yield. This is further aggravated by the fact that the abstraction cost is also artificially low given the 2/3 subsidy of electricity prices. Since aquifers are an open access resource, farmers have no incentives to save water. They are rather interested in increasing production to maximize profit. There is no warranty that the saved volume of water will remain available to them given the absence of ownership, the absence of a pricing scheme, and low abstraction cost. Farmers equipped with modern irrigation systems tend to crop the area irrigated even in summer, which is only possible, under a hot climate, with a modern irrigation system given the very high summer temperatures. The data shown here is evidence that the investment in modern irrigation systems alone is not sufficient to stop the over-abstraction of groundwater. A proper monitoring of groundwater abstraction via the allocation of water quotas should be put in place to ensure the sustainability of agriculture in the future. Only after allocation of quotas can the modern irrigation systems lead to tangible water saving. The attribution of subsidy for the adoption of modern irrigation systems, thus, should be linked to real water saving if monitoring of the aquifers is implemented.

6.3.6 Research

The main research in the water sector is undertaken in Sultan Qaboos University and the research center under MAF with either internal funding or national funding from The Research Center, MAF, and MRMWR. Very limited and sporadic international funding through the Network of Water Research Centers was obtained. The main areas of research are on groundwater hydrogeology, groundwater management and governance, treated wastewater technology and reuse, irrigation technology, and efficiency. Some intents of research on smart irrigation have been observed but still not in the phase of transfer to the farm level. More applied research to the needs of the sector should be funded to help devise optimal solutions to the challenges faced by the water sector. Continuity of funding is a major requirement to bring both institutional and technological innovations.

6.3.7 Food Security Versus Virtual Water/Food Imports

The increase of groundwater salinity and depletion of aquifers are the major threats to the agricultural production (Zekri 2008, 2009; MAF 2012). The cropping pattern in Oman is shown in Fig. 6.3. Fruit trees such as dates, banana, lime, and mango cover about 53% of the total cropped area. Date palms occupy 80% of the area under fruit trees. More than 95% of the fruit trees are of local varieties with poor quality and low yield. Current cereal production is very limited for a niche market. Most vegetables are consumed locally. Export of vegetables is limited to the United Arab Emirates or to a niche market in Japan by a few large vegetable producing companies. Oman meets 80% of its food demand by import, with 100% of rice and 95% of wheat imported (FAO 2016) and achieved 78% self-sufficiency in vegetables and 42% in fruits. Figure 6.4 shows the rate of self-sufficiency for 2016. The Sultanate is expected to depend more on imports in the future as the water resources are getting scarcer and the population is increasing at a high rate.

6.3.8 Water Salinity/Other Pollution Problems

Groundwater pollutants can be classified as chemical and biological pollutants. The most known biological pollutant is the Coliform bacteria, which is attributed to animal and human waste. Several aquifers used for domestic purposes in remote rural areas are affected by biological pollutants due to the unsafe use of sceptic

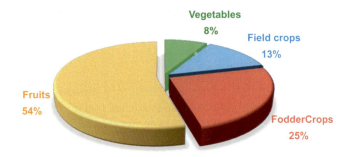

Fig. 6.3 Cropping Pattern. (Source: MAF 2013)

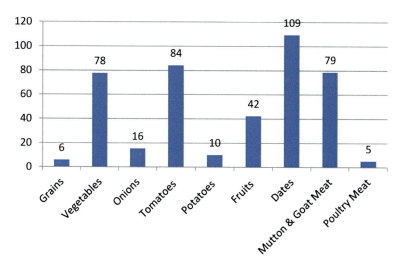

Fig. 6.4 Self-sufficiency ratio of major agricultural products. (Source: FAO 2016)

tanks. Organic pollutants are those related to hydrocarbon chemicals. Oil companies are required to treat all the oil-produced water before injecting it to the ground again or disposing on the surface. The use of fertilizers and pesticides by farmers percolates and reaches groundwater. However, the main source of pollution of coastal aquifers in Oman is the seawater intrusion.

6.4 Urban Water

The two main sources of urban water in Oman are the desalination water provided by large desalination plants and covers 80% of total needs of potable water and public wells, which are connected to the main transmission system network and cover 15%. Other wells such as single or small groups of wells supply water for the scattered houses particularly in rural areas. The remaining 5% of total potable water is covered by small desalination plants and small local distribution network system (PAEW 2016). Currently Oman produces around 1 Mm^3/day of desalinated seawater.

To cover the increased demand of water the government relies on a supply policy based on increasing the number of the large desalination plants or expanding the capacity of existing ones. Earlier the small desalination plants were owned by the Public Authority for Electricity and Water and the electricity was produced at the same site. Recently the Public Authority for Electricity and Water adopted a new policy of "Build Own Operate" in which the desalinated water is purchased from the private sector, which resulted in substantial savings (PAEW 2016). The groundwater extracted from wells is used during peak demand periods and covers 15% of the total water in the large transmission networks (PAEW 2016).

The fundamental element to improve the efficiency of the distribution system is the control of the 30% unaccounted for water or nonrevenue water. Water enters the system from several sources, wells and desalination plants, and is then transferred and distributed through pipe network. Recently, the PAEW installed pressure mitigation valves to regulate the pressure in the network expecting substantial reduction in the leakages.

One major problem related to urban water in countries heavily dependent on desalination is water security. Desalination plants are exposed to risks when seawater intake is polluted by harmful algae bloom, oil spills, or by mechanical failures. One of the options to improve water security is to inject desalinated water in excess of supply during winter low demand periods in close by aquifers. Zekri et al. (2019) estimated that up to 8.4 Mm^3/year of excess winter water can be stored and would serve for emergency as well as pick demand period. The expected potential net benefit of storage and recovery is around $ 17.80 million/year.

6.5 Water and the Environment

The availability of fresh water relies on the function of healthy ecosystems while water balance through water cycle is essential to achieving sustainable water management. Oman produces 4% of the global seawater desalination capacity (Lattemann and Hooepner 2008). The desalination of seawater provides socioeconomic and environmental benefits by supplying unlimited and sustained high-quality potable water, but at the same time, multiple concerns arise as a result of the disposal of the concentrate and chemicals. The discharge of these chemicals and reject brine can lead to water quality degradation, affect the marine system, and increase air pollution through greenhouse gas emissions. Therefore, an integrated management on a regional scale is a necessity to regulate the use of water resources and the various desalination technologies in a sustainable way.

The Ministerial Decision 159/2005 regulates the discharges of reject brine from coastal desalination plants (MRMEWR 2005). Oman has adopted different types of brine discharge, including nonenvironmentally friendly methods such as discharge to infiltration manhole units and surface discharge or lined and unlined evaporation ponds. Desalination plants, which are mainly located in coastal areas, rely on discharging brine to the ocean. The maximum capacity of marine outfall system is estimated at 122,000 m^3/h in order to mix discharge brine with the cooling water from the power generation plants. The brine is usually disposed of with outfall pipes in the sea of a length ranging between 650 and 1200 m. In some of the plants, multiport diffusers are installed as an engineering solution to rapidly dilute the discharged brine (Al-Barwani and Purnama 2011; Purnama 2011). In general, the temperature and concentration of brine are within the Omani regulations and regulatory mixing zone of 150 m radius from the outfall, which is below the maximum permissible limits set (above the surroundings by 1 °C and 2 ppt) (Bleninger and Jirka 2010). However, some damage to large coral reef areas at the bottom of the sea, which was close to the outflow, is observed in Sur where brine disposal is undertaken through an open sea pipe near to the shoreline (Bashitialshaaer 2016).

6.6 Energy–Water Nexus

Water and energy are intimately linked. So far, fossil fuels (natural gas and oil) are the unique source of energy for desalination and water operations in Oman as well as groundwater abstraction and irrigation. While depending on fossil fuel to operate the desalination plants, a loss of opportunity for revenue from exporting hydrocarbon fuel is observed. The analysis of natural gas used between 2000 and 2011 shows that power and water production uses almost 1/5 of the total uses as shown in Fig. 6.5.

The Ministry of Oil and Gas provides natural gas for power generation and associated water production. The introduction of new efficient power plants enhanced the efficiency of natural gas use for water and power production. As a result, the annual amount of gas utilized has decreased at a rate of 4.8% during the period 2010–2016 (OPWP 2016) while the volumes of desalinated water kept increasing, as shown in Figs. 6.6 and 6.7. This is a good indicator of a better management of the

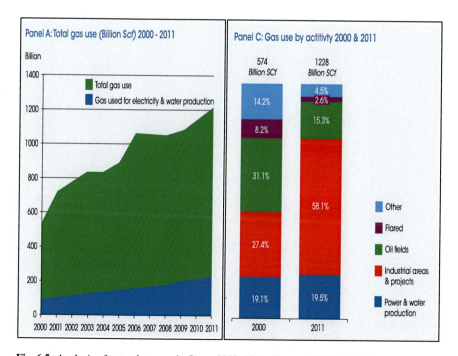

Fig. 6.5 Analysis of natural gas use in Oman 2000–2011. (Source: IRENA 2014)

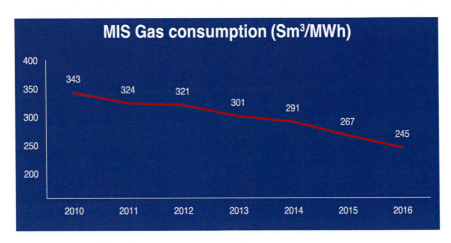

Fig. 6.6 Gas consumption for energy production decrease from 2010 to 2016. (Source: OPWP annual report 2016)

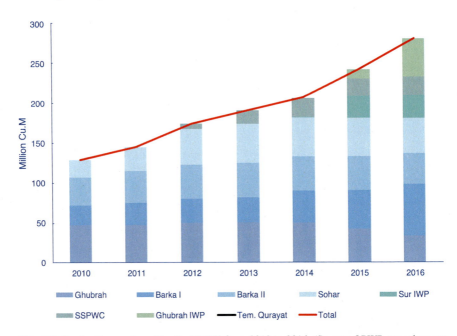

Fig. 6.7 Freshwater purchased by the PAEW from 2010 to 2016. (Source: OPWP annual report 2016)

nonrenewable resources. Oman's natural gas recoverable reserves are estimated at 20 years according to the World Energy Council. Oman is currently a net exporter of natural gas. However, given the increased domestic demand, the country decided to divert gas exports back toward domestic consumption by 2024 (https://www.worldenergy.org/data/resources/country/oman/gas/, 2018).

The move toward the use of renewable energy for water and energy production is on the agenda. Even though Oman meets substantial challenges related to water and energy security it can still use the current wealth of fossil fuels to move forward the essential transitions to assure long-term security of water, energy, and food. Oman plans to reach 10% of its energy from renewable energy by 2020 (IRENA 2015). OPWP expects solar energy projects to supplement hydrocarbon fuel generation in the near future. In the short run, nonrenewable energy is the only source to develop new capacity of power and desalination to meet the peak demand (International Water Summit 2018). It is expected that the introduction of renewable sources of energy will allow a potential reduction of 8% in water use for power generation in Oman by 2030 (IRENA 2015).

6.7 Conclusions

A collective of governmental and nongovernmental administrations are responsible for regulating the water sector in Oman, with each organization being liable for specific tasks. Although the responsibilities are scattered among these entities, often the concerned organizations are consulted and their members are called to sit in committees and participate in the decision-making process. The expected further development of E-government will provide better access and exchange of information among ministries and organizations, which will likely result in improved coordination and sounder decisions. However, given the "private" type of management of some of the water utilities (PAEW, Haya) involved in the sector, often each organization is looking for its own interest that does not necessarily coincide with the national interest and the economic benefit rather than the financial benefit.

Oman as an arid country faces real water challenges given that already the current demand exceeds the supply. The agricultural sector depends totally on groundwater, while natural groundwater recharge rate is very low. The excessive uncontrolled abstraction of groundwater threatens the future generations and causes groundwater salinization through the intrusion of seawater. Around 83% of groundwater is used for crop irrigation. Salinity and water scarcity are the major threats for the agricultural sector. Oman has achieved 78% self-sufficiency in vegetables and 42% in fruits and meets 80% of its food demands by import. The Sultanate's dependence on food imports is expected to increase considerably in the future, given the high population growth (due partly to immigration) and the very limited water resources that do not allow any expansion in irrigation. However, irrigation efficiency and better farmers' organization can increase the rates of self-sufficiency in vegetables and food balance.

The traditional Aflaj systems play a significant role in irrigation efficiency. The water turn depends on timeshare formal water rights. The active millenarian water markets in Aflaj organize the transfer of water to the users who are willing to pay higher prices than others and thus efficiency is guaranteed. In times of drought the markets allow the plantations to be given priority while producers of other annual crops get compensated partially through the market if they sell their water shares. With the presence of the traditional water markets, drought periods are managed smoothly and without conflicts. Besides, the horizontal channels of Aflaj do not put any stress on the aquifers and thus farmers accommodate with the renewable part of the groundwater.

Treated wastewater can become an alternative source for irrigation if the required policy changes of groundwater quota allocation, cost recovery of the treatment, and establishment are implemented.

Desalination plants provide most of the water for urban purposes. The main concern related to desalination is the high cost of freshwater production and the disposal of brine by an environmentally friendly method. Desalinated water supply is expected to increase to cover the rapid increase in urban demand. All desalination plants in Oman are operated using natural gas. Currently, water desalination and power generation use around 20% of total natural gas. In the long run, the authorities are planning for solar energy projects to supplement hydrocarbon fuel generation.

To face the challenges, several reforms are needed and are multifaceted. The top 10 required reforms are (1) to allocate groundwater quotas by user capitalizing from the traditional Falaj knowledge; (2) to professionalize the agricultural sector by requiring qualifications for the expatriate workers in line with what is applied in the health sector; (3) to open up international markets for Omani vegetables that are produced during the winter season; (4) to improve the irrigation efficiency at farm level and explore the introduction of smart irrigation; (5) to encourage the production of vegetables instead of the high water demanding crops such as forage and dates, which can be achieved by banning the production of forages at least in the coastal areas; (6) to stop export of forages and dates from Oman, which will allow saving 60 Mm^3/year; (7) to improve urban water security by storing excess desalinated water for emergency and peak demand periods; (8) to increase domestic water prices to reduce waste and encourage water saving technologies and reuse of graywater for outdoor purposes; (9) to achieve cost recovery of treated wastewater; and (10) to recharge some of the aquifers with treated wastewater instead of losing to the sea.

References

Al-Barwani, H. H., & Purnama, A. (2011). A computational model study of brine discharges from seawater desalination plants at Barka. In R. Y. Ning (Ed.), *Expanding issues in desalination.* Rijeka: InTech Open Access Publisher.

Al-Maktoumi, A., El-Rawy, M., & Zekri, S. (2016). Management options for a multipurpose coastal aquifer in Oman. *Arabian Journal of Geosciences, 9*(14). https://doi.org/10.1007/s12517-016-2661-x.

Baerenklau, K., Schwabe, K., & Dinar, A. (2014). Allocation-based water pricing promotes conservation while keeping user costs low. *ARE Update, 17*(6), 1–4. University of California. Giannini Foundation of Agricultural Economics.

Bashitialshaaer, R. (2016, January/February). Desalination report for Omani public Authority for Electricity and Water (PAEW), six weeks in Oman. *International Desalination Association IDA-NEWSLETTER, 25*(1), 7–8.

Bleninger, T., & Jirka, G. H. (2010). Final report on environmental planning, prediction and Management of Brine Discharges from desalination plants, Middle East desalination research center Muscat, Sultanate of Oman, MEDRC series of R&D reports, MEDRC project: 07-AS-003.

El-Rawy, M., Al-Maktoumi, A., Zekri, S., & Al-Abri, R. (2018). Hydrological and economical feasibility of mitigating a stressed coastal aquifer using managed aquifer recharge: A case study of Jamma aquifer, Oman. *J Arid Land (2019), 11*(1), 148–159. https://doi.org/10.1007/s40333-019-0093-7.

FAO. (2008). Policy options and alternatives for the cultivation of fodder crops in Al-Batinah Region Sultanate of Oman. Final Report. 180 pages.

FAO. (2016). www.faostat.fao.org.

FAO-MAF. (2016). *Sustainable agriculture and rural development strategy towards 2040*. 133 pages.

Haya. (2018). *Haya water treated effluent utilization strategy*. Oman Water & Wastewater Conference, 30th April–2 May 2018. OmanExpo, Muscat, Sultanate of Oman.

International Water Summit. (2018). *Energy efficient desalination, meeting the GCC's water needs in an environmentally sustainable way*. Available online: https://www.internationalwatersummit.com/_media/Energy-Efficient-Desalination-2018.pdf

IRENA, International Renewable Energy Agency. (2014). *Sultanate of Oman a renewables readiness assessment*. Available online: https://www.paew.gov.om/PublicationsDoc/RE-Renwable-Radiness-Assessment

IRENA, International Renewable Energy Agency. (2015). *Renewable energy in the water, energy and food Nexus*. Available online: http://www.irena.org/documentdownloads/publications/irena_water energy_food_nexus_2015.pdf

Joseph, Sokina. (2017). *The Sultanate of Oman's agricultural and domestic sector: Using water footprint and virtual water tools to quantify and assess impacts of water use*. MSc Thesis, Imperial College London, Faculty of Natural Sciences, Centre for Environment Policy.

Kotagama, H., Zekri, S., Al Harthi, R., & Boughanmi, H. (2016). Demand function estimate for residential water in Oman. *International Journal of WaterResources Development*. https://doi.org/10.1080/07900627.2016.1238342.

Lattemann, S., & Hoepner, T. (2008). Impact of desalination plants on marine coastal water quality. In Barth, Abuzinada, Krupp, & Böer (Eds.), *Marine pollution in the gulf environment*. Dordrecht: Kluwer Academic Publishers.

MAF. (2004). *Ministry of agriculture and fisheries, annual book of agriculture statistics*.

MAF. (2012). *Ministry of Agriculture and fisheries (MAF)*. Sultanate of Oman, International Center for Biosaline Agriculture (ICBA) Dubai, UAE, "Oman salinity strategy: Assessment of Salinity Problem".

MAF. (2013). *Ministry of Agriculture and fisheries*. Agricultural Census.

MRMWR, Ministry of Regional Municipalities and Water Resources, Water Resources in Oman (2008). Available online: http://www.omanws.org.om/images/publications/5465_Water_Atlas_E.pdf

MRMWR. (2013). *Ministry of Regional Municipalities and Water Resources*. Sultanate of Oman. http://www.mrmwr.gov.om

MRMEWR. (2005). *Ministerial decision no: 159/2005 promulgating the bylaws to discharge liquid waste in the maarine environment* (p. 10).

NCSI. (2017). National Center for statistics and information. Statistical year book, issue 45, August 2017.

OPWP. (2016). *Oman Power and Water Procurement Company, annual report*. Available online: http://www.omanpwp.com/PDF/AnnualReport2016.pdf

OPWP, OPWP's 7-Year Statement. (2016–2022). Available online: http://www.omanpwp.com/PDF/7YS%202016-2022%20Final%20.pdf

PAEW. (2016). *Public Authority for Electricity and Water.* Annual report. Available online: https://www.paew.gov.om/getattachment/1340e690-b826-4510-a83d-ed9b55b69a6c/

Purnama, A. (2011). Analytical model for brine discharges from a sea outfall with multiport diffusers, proc. International conference on environmental systems engineering and technology, Paris, France, August 24–26, 2011 (World Academy of Science, Engineering and Technology, Volume 80, August 2011, pp 57–61).

Purnama, A., Al-Barwani, H. H., Bleninger, T., & Doneker, R. L. (2011). CORMIX simulations of brine discharges from Barka plants, Oman. *Desalination and Water Treatment, 32*(1–3), 329–338. https://doi.org/10.5004/dwt.2011.2718.

Wichelns, D. (2014). Do estimates of water productivity enhance understanding of farm-level water management? *Water, 2014*(6), 778–795. https://doi.org/10.3390/w6040778.

Zekri, S. (2008). Using economic incentives and regulations to reduce seawater intrusion in the Batinah coastal area of Oman. *Agricultural Water Management, 95*(3), 243–252. https://doi.org/10.1016/j.agwat.2007.10.006.

Zekri, S. (2009). Controlling groundwater abstraction online. *Journal of Environmental Management, 90*(11), 3581–3588. https://doi.org/10.1016/j.jenvman.2009.06.019.

Zekri, S., & Al-Marshudi, A. S. (2008). A millenarian water rights system and water markets in Oman. *Water International, 33*(3), 350–360. https://doi.org/10.1080/02508060802256120.

Zekri, S., Kotagama, H., & Boughanmi, H. (2006). Temporary water markets in Oman. *Agricultural and Marine Sciences, 11*(SI), 77–84.

Zekri, S., Fouzai, A., Naifer, A., & Helmi, T. (2012). Damage cost in dry Aflaj in the Sultanate of Oman. *Agricultural and Marine Sciences, 17*, 9–19.

Zekri, S., Al Harthi, S., Kotagama, H., & Bose, S. (2016). An estimate of the willingness to pay for treated wastewater for irrigation in Oman. *Journal of Agricultural and Marine Sciences, 21*(1), 57–63. https://doi.org/10.24200/jams.vol21iss0pp57-64.

Zekri, S., Madani, K., Bazargan-Lari, M., Kotagama, H., & Kalbus, E. (2017, February). Feasibility of adopting smart water meters in aquifer management: An integrated hydro-economic analysis. *Agricultural Water Management, 181*, 85–93. https://doi.org/10.1016/j.agwat.2016.11.022

Zekri, S., Triki, C., Al-Maktoumi, A., Bazargan-Lari, M. (2019). Optimal storage and recovery of surplus desalinated water. 8th ICWRAE International Conference on Water Resources and Arid Environments. 22–24 January 2019. King Saud University. Riyadh. Saudi Arabia.

Dr. Slim Zekri is Professor and Head of the Department of Natural Resource Economics at Sultan Qaboos University (SQU) in Oman. He earned his PhD in Agricultural Economics and Quantitative Methods from the University of Cordoba, Spain. He is Associate Editor of the journal *Water Economics and Policy.* He has worked as a Consultant for a range of national and international agencies on natural resource economics, policy and governance, agriculture, and water economics in the Middle East and North Africa. He is Member of the Scientific Advisory Group of the FAO's Globally Interesting Agricultural Heritage Systems. His main research interests are water economics and environmental economics. In 2017, he was awarded the Research and Innovation Award in Water Science from the Sultan Qaboos Center for Culture and Science.

Chapter 7
Water Resources in the Kingdom of Saudi Arabia: Challenges and Strategies for Improvement

Mirza Barjees Baig, Yahya Alotibi, Gary S. Straquadine, and Abed Alataway

Abstract Saudi Arabia is an arid country that lacks permanent water-bodies. Saudi Arabia relies on its oil resources to operate its desalination plants to supply potable water. Paradoxically, it has the third highest per capita fresh-water consumption in the world, despite being one of the world's driest countries. Extensive agricultural programs almost depleted the nonrenewable groundwater and deteriorated water quality. The rates of water being used by the urban population and the agricultural sector for producing crops appear to be wasteful (MEWA 2018). Efforts have been made to develop an extensive but efficient water transmission system. The water sector's infrastructure is outdated and, in some areas, deteriorating. It is estimated that an average of 20% of distributed water remains unaccounted. The Saudi Government has streamlined the water sector's regulatory regime through implementation of a strategic water policy. Through its water policy, it is looking at the most efficient ways to produce and regulate water. No country can be expected to try every possible solution, but in Saudi Arabia, no single solution would be adequate to address this complex issue. However, the most important step would be to bring behavior change in the society to use water economically and wisely by adopting water conservation practices.

Keywords Water resources · Ruthless extraction · Water prices · Low tariff · Sustainable agriculture · Water policy

---

M. B. Baig (✉) · Y. Alotibi
Department of Agricultural Extension and Rural Society, College of Food and Agriculture Sciences, King Saud University, Riyadh, Saudi Arabia
e-mail: mbbaig@ksu.edu.sa; khodran@ksu.edu.sa

G. S. Straquadine
Utah State University, Logan, UT, USA
e-mail: gary.straquadine@usu.edu

A. Alataway
Prince Sultan Institute for Environmental, Water and Desert Research, King Saud University, KSA, Riyadh, Saudi Arabia
e-mail: aalataway@ksu.edu.sa

© Springer Nature Switzerland AG 2020
S. Zekri (ed.), *Water Policies in MENA Countries*, Global Issues in Water Policy 23, https://doi.org/10.1007/978-3-030-29274-4_7

7.1 Introduction

The Kingdom of Saudi Arabia (KSA), covering an area of 2.25 million km^2 with limited fresh water supplies, is an arid and water-deficit country. It lacks perennial rivers or permanent water bodies. Characterized by low rainfall and high evaporation rates, KSA is one of the driest areas of the world. Due to the discovery of fossil-fuel reserves, KSA has witnessed remarkable economic development in the past four-decades resulting in elevated living standards of its citizens. Increasingly, KSA has experienced increased population and mass-scale migration to the urban areas in search of better opportunities. The population of Saudi Arabia increased from about 4 million in 1960 to about 32.5 million in 2018. With the present population, KSA is the 41st most populous country in the world (General Authority for Statistics 2016). The population is expected to reach 34.4 million by 2020 and it is predicted to grow by 77% to more than 56 million by 2050 (Rambo et al. 2017; Central Statistics Agency Statistics 2017). These developments have placed extraordinary pressure and increased demand on the country's limited water resources. In order to meet the needs of its citizens, the limited natural renewable water resources in the country have been heavily overexploited. Aquifers are the only source of natural water in the country. The per capita average daily use of water in Saudi Arabia has been increasing since 2009 when it hit 227 l/day and recorded a gradual increase to touch 270 l/day in 2016 – the 3rd highest in the world (GASK 2016).

In the 1980s, the Kingdom started its very ambitious agricultural program to realize self-sufficiency to meet its food requirements. The goals were successfully achieved and the kingdom witnessed self-sufficiency in many food commodities and became the 6th largest wheat exporter at the expense of its water resources. Extensive groundwater extractions from the nonrenewable aquifers for the last few decades resulted in substantial declines in groundwater levels and deterioration of groundwater quality. Behaviors such as liberal water use, careless water supply management, unchecked population growth, and irresponsible agricultural policies developed an unsustainable culture of water usage in Saudi Arabia (Lippman 2014). Urbanization, rising living standards, a continued influx of expatriates, expansion of agriculture and development of new housing, shopping malls, etc. have also taxed water resources (FAO 1997, 2009; Baig and Straquadine 2014).

The Kingdom has a limited reserve of nonrenewable groundwater as well as low recharge rates (National Water Strategy 2016). Despite the fact that renewable water sources in Saudi Arabia are very limited, KSA consumes nearly 7 billion cubic meters of water daily, 60% of which is desalinated. Of this, 40% comes from public desalination stations and 20% from stations operated by private sector (Al-Suhaimy 2013).

Presently, the water sector of the Kingdom is facing several challenges that threaten water and food security, energy security, and development as a whole. Fuel subsidies and desalinated water deplete the energy resources along with consequent environmental cost, low water tariffs, and the increased leakage of water supply network. All of these factors increased the cost of drinking water, the cost of safe disposal of wastewater, the pressure on desalination plants and sewage treatment plants, as well as air pollution from desalination plants (The State of the Environment 2017). The scarcity of renewable water resources and the ever-increasing water demand by the society and various economic sectors are the major challenges faced by KSA to realize sustainable development.

The purpose of this chapter is to review the present status of water resources and provide information on demand, supply, and policies. The chapter identifies problems and challenges, and suggests initiatives and measures to improve the sustainability of water resources of the Kingdom.

7.2 An Overview of the Water Resources in the Kingdom

With a semi-arid environment, the Kingdom is exposed to temperature variability, low annual rainfall, no natural perennial flow, and limited groundwater reserves (Chowdhury and Al-Zahrani 2013). Saudi Arabia has no permanent lakes, perennial surface watercourses, or rivers. The KSA experiences high evaporation rates and on an average less than 100 mm annual rainfall, which limits the availability of surface water sources and supplies. Due to limited rainfall and excessive consumption, the major groundwater aquifers are being depleted. Total fresh water produced in Saudi Arabia increased from 17.5 million cubic meters in 2010 to about 24 million cubic meters in 2016. Despite the increase of desalinated water, the percentages by different sources (i.e., surface (1%), groundwater (91%); desalination (8%), and wastewater treatment (1%)) remained about the same from 2010 to 2016 (GSA 2018). Water produced from various resources with their share in the Kingdom is depicted in Fig. 7.1.

Water resources in the Kingdom are placed in two groups. Traditional sources include renewable surface water and renewable and nonrenewable groundwater, and nontraditional water sources include desalinated and treated wastewater. The four main water resources in Saudi Arabia are as follows: surface water, groundwater (nonrenewable groundwater from the deep fossil aquifers and renewable groundwater from shallow alluvial), desalinated water, and reclaimed wastewater (treated).

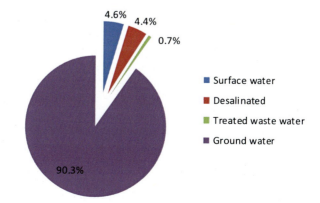

Fig. 7.1 Water resources with their share in the Kingdom. (Source: Global water Market 2017: Meeting the world's water and wastewater needs until 2020. Volume 4: Middle East and Africa, Saudi Arabia)

7.2.1 Surface Water and Runoff Collection in Dams

Surface water provides 10% of water supply, mainly in the western and southwestern regions of the Kingdom. Rains are not frequent but intense; intermittent flash floods (Wadis) and random flooding due to rains generate surface water. Water collected through rainfall (surface water) is stored in dams (World Bank 2005) and is also often used for drinking purposes. Dams trap surface water runoff during rainstorms, occurring in the coastal areas and southwest highlands. Surface water is replenished by the flash floods, experienced from November to April (Lovelle 2015). The Kingdom has taken initiatives like construction of new dams and development of a Rainwater Harvesting (RWH) System. The largest quantity of runoff (about 60%) from rainfall occurs in the southwestern region of the Kingdom. According to experts, this water could be collected, stored, and transported to other parts of the country through new and existing pipelines (Chowdary and Alzahrani 2012).

Runoff quantity and duration are temporarily and spatially varied depending upon the rainfall pattern and distribution. In 2013, an estimated volume of 0.6 billion m^3 was captured from intermittent flash floods by the 260 irrigation dams (Ouda 2013a). In 2015, approximately 1.4 billion cubic meters (BCM)/year of runoff were collected by 302 dams (Chowdhary and Al-Zahrani 2015). In 2016, there were 449 dams to collect, store, and recharge runoff and to control flash floods (MEWA 2017; Tarawneh and Chowdhury 2018). According to the most recent estimates reported by the National Water Strategy (2016), total reserves of usable water from dams are about 1.6 billion cubic meters annually. About 73% of the total usable water comes from dams in Asir, Makkah, and Jazan. These areas contain an abundance of renewable groundwater and surface water due to their topography and nonporous rocky nature at the Arabian Shield (National Water Strategy 2016). The Kingdom currently has 535 operational dams, with a storage capacity of over 2 billion cubic meters and more dams are under construction to enhance water storage and drinking water resources (Argaam 2018).

7.2.2 Groundwater Resources

Groundwater comes from two sources. The first source is nonrenewable groundwater from the deep fossil aquifers. The second source is renewable groundwater from shallow alluvial aquifers. Combined groundwater sources accounts for 40% of supply. Groundwater is available in the KSA from the shallow renewable and nonrenewable groundwater aquifers to use for different proposes (The State of the Environment 2017). Groundwater resources are the prime source of available water in the Kingdom. More than 80–90% of the national water use is satisfied from groundwater which is pumped from local aquifers (Al-Salamah et al. 2011). The groundwater is stored in several aquifers in Saudi Arabia with minimum annual recharge. It is estimated that the total groundwater storage in the Arabian Peninsula is 80,000 km^3 (Abderrahman and Rasheeduddin 2001) and the total groundwater reserve in Saudi Arabia is estimated to be 2259 billion cubic meters (Abderrahman and Al-Harazin 2008).

7.2.2.1 Nonrenewable Groundwater Resources in the Arabian Shield

The Kingdom has a reserve of nonrenewable groundwater that spreads across a range of layers. Water and groundwater are collected in more than 20 primary and secondary aquifers prevailing in many areas of the Kingdom. All basic aquifers are located on the Arab shelf; the Arabian Shield contains many primary or secondary aquifers because of its nonporous rocky nature (National Water Strategy 2016). Nonrenewable groundwater sources, shallow alluvial and deep rock aquifers are the two paramount sources of groundwater. The deep rock aquifers are sedimentary in origin and hold "fossil" water (DeNicola et al. 2015).

The groundwater in the deep sandstone aquifers is nonrenewable or "fossil" water. This type of water was formed approximately 10–32 thousand years ago and extends over thousands of square kilometers with poor natural recharge through upland and foothill zones where the rocks have surface outcrops (Chowdhury and Al-Zahrani 2015). Tarawneh and Chaudhry (2018) report that the fossil water is available in the sand and limestone formations at the depth from 150 to 1500 m. Nonrenewable groundwater reserves were estimated at 259.1–76.6 billion cubic meters. In contrast, the statistics of Ministry of Environment, Water and Agriculture, (2018) disclose that the nonrenewable groundwater reserves of the Kingdom are estimated at 2360 billion cubic meters. Out of this total, about 1180 billion cubic meters can be extracted. These nonrenewable groundwater reserves are capable of producing 20.6 billion cubic meters annually, covering 90% of agricultural production while covering about 35% of municipal needs and 6% of industrial needs.

Locations of the aquifers in the Arabian Shield have been indicated in the map presented in Fig. 7.2.

However, the rates of extraction from renewable groundwater resources (i.e., shallow aquifers) are higher than the recharge and are getting depleted. It has been estimated that at the current rate of withdrawal, these water supplies will be gone in less than 50 years (Drewes et al. 2012). Most of the water extracted was pumped out of fossil water in the Kingdom. This sort of heavy extraction of water from fossil aquifers poses a serious threat to the security of nonrenewable water resources (Lovelle 2015).

7.2.2.2 Renewable Groundwater Resources

Renewable groundwater is found in the shallow and deep layers and surface water in the valleys in the Kingdom. According to preliminary estimates, renewable groundwater is currently estimated at 2.8 billion cubic meters annually in the Arabian Shield area (National Water Strategy 2016). However, the rates of extrac-

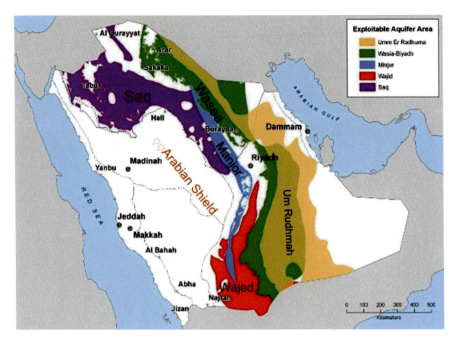

Fig. 7.2 Locations of the aquifers in the Arabian Shield

tion from renewable groundwater resources (i.e., shallow aquifers) are higher than the recharge and are getting depleted. In a recent study conducted for the Gulf Cooperation Council, it was revealed that KSA consumes between 10 and 39 times the amount of renewable water available, depleting its aquifers at a much faster rates than they can be replenished by rainfall.

7.2.3 Desalination Plants

Desalinated water produced by 35 desalination plants provides 50% of the drinking water. The share of desalinated water to the total fresh-water resources ranged from 7–9% during the period 2010–2017 (General Authority for Statistics 2018). In the absence of the perennial rivers and other permanent water bodies and due to the rapidly depleting underground water resources, water produced by the desalination plants seems to be the best, most valid, and the right strategic option to make the drinking water available and meet growing domestic water demand in the country (Al-Ibrahim 2013; Ouda 2014a). The KSA is the largest desalinated water producer in the world and it currently produces about one-fifth of the world's total. The KSA government considers seawater desalination as the strategic option to alleviate the water scarcity problem, and to meet the country's ever-growing domestic water demand (Ouda 2014a).

7 Water Resources in the Kingdom of Saudi Arabia: Challenges and Strategies... 141

Table 7.1 Water produced (thousands of cubic meters) by desalination plants for years (2012–2016)

Plants	2012	2013	2014	2015	2016
Al-Jubail	380,990	388,028	388,614	378,256	308,655
Al-Khobar	144,514	129,703	119,514	128,260	114,057
Ras-Alkhair	—	—	55,169	197,463	256,963
Yambu	122,765	136,185	140,629	132,686	124,369
Jaddah	132,226	163,639	189,200	196,830	174,309
Al-Khafii	8023	8032	8120	8058	6906
Al-Shuibah	170,274	176,699	182,870	193,422	150,544
Al-Shoqiq	15,134	28,576	30,746	31,489	27,627
Others	23,307	24,297	25,102	25,698	22,094
Total	997,233	1,055,159	1,139,964	1,292,162	1,185,524

Source: General Organization for Desalination; General Authority of Statistics, 2017

Saudi Arabia currently has 35 desalination plants located on the east and west coast of the Kingdom. The latest data indicate that the total capacity of desalination reached 6.28 million cubic meters per day in 2015. This capacity is expected to increase to 7 million cubic meters per day by 2020. Water produced by various desalination plants for the years 2012–2016 is depicted by Table 7.1.

The Saline Water Conversion Corporation (SWCC) increased desalination plants' production capacity from about 200 million m^3/year in 1980 (Al-Zahrani and Baig 2011; MWE 2012; SWCC 2014) to about 1.18 billion m^3/year in 2016 and 1.2 billion m^3/year in 2017 (General Organization for Desalination; General Authority of Statistics 2017). In order to make water available to the users, there are 2500 miles of pipelines from these desalination plants across the kingdom (CDSI 2010).

The demands for water and electricity production are growing by 8% every year (NACHET 2015). The consumption of desalinated water is increasing by about 14% annually. This is twice the growth in total domestic water consumption and six times the population growth rate (Rambo et al. 2017). On average, it is twice as high as the global average because it uses much more water than countries with larger water resources (Nachet and Aoun 2015). However, the kingdom plans to double its desalination capacity in the next decade (Rambo et al. 2017). The desalination process requires high energy input. More than half of domestic oil consumption is required to run the desalination plants (Rambo et al. 2017) and the present demand rates suggest that this figure will reach 50% by 2030 (Al-Hussayen 2009).

In addition, the high costs and energy consumption linked with the desalination process have impacted the desalination ability of the Kingdom; therefore, its water security depends upon a stable oil supply (Lovelle 2015). With the key problem identified, it is concluded that desalination is very expensive and may not be sustainable on a long-term basis. It represents 10–20% of energy consumption in the KSA (Rambo et al. 2017).

It has been noted that the cost per unit of water production is relatively high because of high transport cost of pumping water from the coasts into the interior of the Kingdom. Presently, desalination plants employ three types of techniques: (1)

multi-stage flash distillation, (2) reverse osmosis, and (3) multiple distillations. Multi-stage flash distillation technology is currently the predominant technology being employed in the desalination plants in the Kingdom, accounting for 62% of the total existing production capacity (National Water Strategy 2016).

Quoting the Minister of Environment, Water and Agriculture report (Al-Awsat 2018), "nine new water desalination plants using modern techniques on the Red Sea coast would be built in the 2019–20 to improve the quality of water services." With the value of $533 million and equipped with the modern technologies, these plants will have the capacity to produce a volume of 240,000 cubic meters per day. The world's biggest desalination plant in Ras Al Khair Industrial City, 75 km north-west of Jubail, was built at a cost of $7.2bn. The plant has a daily production capacity of 1.025 million cubic meters of desalinated water and 2600 MW of electricity (Arabian Business 2014b). Another giant desalination plant was built in Rabigh on the Red Sea near Taif with a capacity of 600,000 cubic meters (158 million gallons) of water a day upon its completion in 2018 (Arabian Business 2014a). The capital, Riyadh, meets most of its water requirements from the desalinated water pumped from the Arabian Gulf along 467 km into the city.

7.2.4 Treated Wastewater

Reclaimed wastewater (treated) provides one percent of the water supply. Treated water is safe to be used for industrial processes, refrigeration, and agricultural purposes. Treated water is an important resource in a struggling country of water scarcity and should be taken into account in the supply system. The total amount of treated wastewater reused throughout the Kingdom in 2015 was 0.61 million cubic meters (MCM) per day; about 0.40 million m^3/day is used in agriculture. On average, 17% of wastewater is reused (National Water Strategy 2016). The wastewater collection system needs to be improved; coverage is still relatively low (60% by 2015). In addition, existing wastewater treatment plants operate at a high level of use, resulting in reduced quality of treated water (National Water Strategy 2016). The use of recycled wastewater lessens the dependence and reduces the pressures on the fresh water resources. It also reduces the amount of treated and nontreated effluent into the environment. These effluents deposit organic and inorganic nutrients (e.g., nitrogen and phosphate) into water systems, which can cause eutrophication and algal blooms and severely degrade existing bodies of water (Toze 2004). Water reuse and recycling are also viewed as a positive step toward climate change adaptation and mitigation.

In 2012, an area of about nine thousand hectares of date palms and forage crops near Riyadh was irrigated using about 146 MCM of treated wastewater. Wastewater is also reused for irrigating landscape plants, trees, and grass in municipal parks in several cities, such as Dhahran, Jeddah, Jubail, Riyadh, and Taif (Ouda 2014b). The use of recycled wastewater in agriculture saves energy and reduces the cost of freshwater pumping, providing irrigation and reducing the water footprint of food pro-

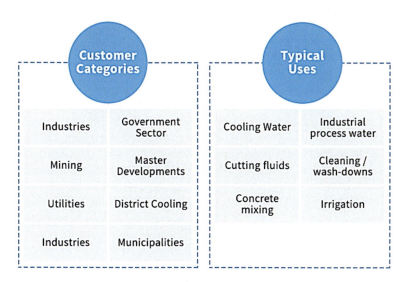

Fig. 7.3 TSE customer categories and uses. (Source: NWC 2018)

duction. It also can provide adequate nutrients and fertilizer for crops so that mining for mineral fertilizers can be decreased. For example, it has been demonstrated that reusing treated municipal wastewater for agricultural irrigation in Saudi Arabia provided adequate nutrients, lowered costs for irrigation and fertilization, and increased yield and profit for wheat and alfalfa (Aljaloud 2010).

Saudi Arabia has established numerous centralized and decentralized wastewater treatment plants. The system involves the collection of wastewater from individual homes, home clusters, isolated communities, industries, and institutional facilities (Tchobanoglous 1995). Recently, a new sewage treatment plant producing a tertiary treated effluent with an average capacity of 400,000 m^3/d and the maximum capacity of 640,000 m^3/d was built. The estimated total available supply of treated sewage effluent (TSE) from the six largest cities exceeds 4.8 million m^3/day (NWC 2018). Customer categories and uses of TSE are presented in Fig. 7.3.

The Kingdom plans to provide water supply to all cities with a population of more than 5000 people by 2025 (Drewes et al. 2012) and to invest US$23 billion into sewage collection and treatment infrastructure (Kajenthira et al. 2012). A large amount of treated wastewater was discharged to Hanifa Valley, which is located in eastern Riyadh and extends beyond the city into the surrounding rural areas. The vision for Hanifa Valley is to use treated wastewater to transform this urbanized valley into a ribbon of naturalized parkland to promote the area as a safe, green, and healthy environment. Moreover, this will connect the area with residential development, farming, recreation, cultural activities, and tourism (Argaam 2018). In order to address the water deficit, Saudi Arabia has started an increased utilization of water recycling. Regarding the reuse of water, the KSA is one of the ten largest consuming countries in the world. Looking forward, the country has aggressive long-term goals of increasing water reuse to more than 65% by 2020 and more than

90% by 2040, all by transforming more of its existing and planned wastewater treatment assets into source water suppliers across all sectors (Chowdhury and Al-Zahrani 2013; Saudi Arabia Water Report Q4 2013; Al-Saud 2013; Scotney 2014; Enssle and Freedman 2017).

The city of Riyadh has been very successful in using nearly 50% of its treated wastewater (about 120 million m³). Treated wastewater is currently being used at various industrial and commercial enterprises of the city, which has the potential to expand in the future due to the high cost of desalinated water about USD 0.8/m³ (Ouda 2014a; Saline Water Conversion Corporation 2010, 2012). Treated wastewater is drained through a canal 40 km south of the Al-Kharj area, where it is stored in a pond and then filtrated through the sandy soil to groundwater. The expected growth in wastewater collection and treatment services will produce more treated wastewater (i.e., about 2.5 km³/y in 2035). The priority is expected to shift from the ongoing major agricultural use to industrial use, with higher anticipated revenue. Use of treated wastewater is projected to be greatest in the Riyadh, Makkah, Medina, and Eastern Province regions, which are home to the KSA's six largest cities (KAUST 2011).

7.3 Domestic Water Pricing

Consumers in Saudi Arabia pay only 1% of government's costs for producing desalinated water (Lovelle 2015). The introduction of a water pricing system will not only help meet some production costs, but also increase the value of the water so that consumers better manage this resource. Attaching a higher value to water could lead to a change in behavior in Saudi Arabia, making the citizens adopt sustainable water consumption (Lovelle 2015). The new tariffs and prices for water consumption and sanitation services received in 2016 have been applied and are presented in Table 7.2. For the first block, the price is US$0.027 for the first fifteen cubic meters, US$0.04 for the second block for consumption less than 30 cubic meters, US$0.53 for the third block for demand less than 45 cubic meters, US$1.07 for the forth block for consumption less than 60 cubic meters, and US$1.6 for the consumption above 60 cubic meters. New tariffs and prices for the volume of water consumed and sanitation serviced received have been implemented in 2016 and are presented in Table 7.2.

Table 7.2 New tariffs for water and sanitation implemented in January 2016 in SAR

Quantity (m³/Month)	Potable water	Sanitary wastewater	Total	Invoice amount
Less than 15	0.10	0.05	0.15	2.25
16–30	1.00	0.50	1.50	22.50
31–45	3.00	1.50	4.50	67.50
46–60	4.00	2.00	6.00	90.00
Above 60	6.00	3.00	9.00	SAR 9/m³

Source: National Water Company (NWC 2017); Note: 1 US $ = 3.75 SAR

7.4 Prime Water User Groups and Their Demand in the Kingdom

The International Monetary Fund (IMF) views Saudi Arabia "at risk" as an area for having low water supply and with high demand placed upon the available resources by its agricultural, municipal, and industrial sectors (IMF 2015). The prime consumers in Saudi Arabia are agricultural, municipal, and industrial sectors and their consumption rates are depicted in Fig. 7.4 and their demands for water are shown in Fig. 7.5.

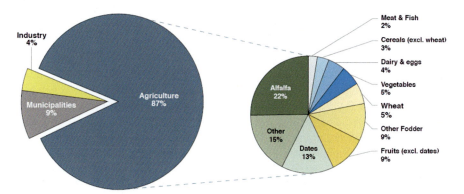

Fig. 7.4 Water consumption by different sectors and various agricultural crops. (Source: Napoli et al. 2016; Image Credit: KAPSARC 2016)

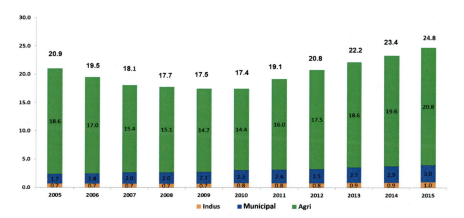

Fig. 7.5 Volume of water consumption in the three main sectors in the Kingdom (billion cubic meters/year) 2005–2015. (Source: Al-Subaiee 2018)

7.4.1 Agricultural Sector

Being the most water-deficit country in the world, its water resources are unable to further support its agricultural needs. In order to achieve food security, the agriculture sector previously received prime importance in the national developmental plans. In a recently published government report, it was noted the absence of tariffs on groundwater for agricultural purposes and the lack of meters on wells led to a quantitative and qualitative deterioration of groundwater and resulted in a threat to the Kingdom's water and food security.

The demand for water for agricultural purposes declined between 2005–2010 due to the phase-out of the wheat program announced by the Government of Saudi Arabia. However, the volume of water consumption in the Kingdom for the period 2011–2015 by agricultural, industrial, and municipal sectors is presented in Fig 7.5, and all sectors are increasing their water consumption.

In order to conserve nonrenewable fossil water resources, the Saudi government adopted a rollback policy in 2008 to discourage domestic wheat production. It rationalized cereal crops and promoted the cultivation of high-return-value vegetables and fruits (Ouda 2014a). The Kingdom stopped buying wheat from local farmers since 2016 and in the future all wheat will be imported. However, the policy to stop wheat farming did not work well and cessation of wheat farming moved farmers toward alfalfa cultivation – a crop that consumes much more water than wheat. This fact prompted the Saudi government in 2015 to stop the cultivation of green fodder within 3 years (Khrais 2016). The farmers were allowed to keep fodder (alfalfa) cultivation (that consumes three times more water than wheat) to supply feed to the dairy industries. Yet, farmers are not allowed to grow fodder after 2019 due to the complete phase-out of the program. However, another scientist (Ouda 2014a) is of the opinion that the new policy resulted in prompt reduction in irrigation water demand and the cultivated area of cereal crops. Some researchers report that areas under cultivation decreased, whereas it increased according to the World Bank – Global Economy Report (2018).

Grain imports for the 2017/18 marketing year are expected to reach 20.5 million tons, an increase of almost 9% over the previous year and 3 million tons over the 5-year average. The import of barley and maize, mainly used for animal feed, accounts for the bulk of grain imports and is expected to reach 8 million tons and 5 million tons, respectively. The government is encouraging the use of processed foods instead of raw barley to reduce barley imports by 1.5 million tons by 2020. Wheat imports are expected to remain high at 3.4 million tons, while rice imports should remain stable at 1.2 million tons (GIEWS 2018).

7.4.2 Domestic Water

The average per capita water use in the Kingdom was 266 l per day (96.8 cubic meters per year) in 2017 compared to 270 l per day (98.4 cubic meters per year) in 2016, recording a decline of 1.7% for the first time since 2013 (General Statistics Authority 2018). The KSA has one of the world's highest per capita water consumption. The average water consumption per capita is estimated at 100–350 l per day for urban areas and between 15 and 20 l for rural areas. Despite the desert landscape in the Kingdom, nearly 97% of the population has access to drinking water, as stated in the World Food Security Index 2015 (Lovelle 2015).

7.4.3 Industrial Water Use

The Kingdom experienced an increased water demand by its industrial sector from 56 to 713 MCM/year from 1980 to 2009 (Chowdhry and Al-Zahrani 2015). Due to an impressive and rising growth of the industrial sector over the past several decades, the water demand for the sector has increased at a rate of 7.5% per year. As Fig. 7.5 reveals, the sector consumed almost 1.0 BCM in 2015. It is predicted that industrial water use would continue to grow by around 50% in the next 15 years. As the Kingdom plans to diversify its economic base away from its dependence on oil and natural gas, in turn the industrial and manufacturing sectors are attaining a prominent place in the recently launched Vision 2030; the situation would further put pressure on the water resources of the Kingdom (Stratfor 2017).

7.5 The Water Sector Faces Many Challenges (Table 7.3)

Table 7.3 An overview of the challenges facing the water sector in the Kingdom

Water scarcity and poor management efficiency and development, with high consumption of the agricultural sector, which affected the sustainability of nonrenewable water resources
Increasing urban water consumption as a result of population and economic growth and dependence on energy-intensive sources, placing economic and environmental burdens on the sector
Low level of service and sector readiness for risk management due to scarcity of qualified human capital and inadequate planning, operation, and customer service
Lack of financial management and poor work on the commercial bases in the sector, which affects transparency and economic sustainability
Absence of institutional and regulatory reforms affecting the effectiveness of water sector governance and its contribution to increasing the participation of the private sector in the kingdom

Al-Subaiee 2018; Ministry of Environment, Water and Agriculture (MEWA) 2018

7.6 Prime Constraints Faced by the Water Sector

7.6.1 Desalination Plants Have Significant Environmental Costs

Despite the productive role and technological achievements of desalination plants in producing drinking water, they come with significant environmental costs. Desalination plants produce carbon emissions by burning fuel and diverting salt from the process back to the sea. This has increased the temperature and salinity of the Gulf, with the latter increasing nearly 2 percent in the last 20 years, having adverse effects on marine life and the ecosystems. Increasing the salt content makes future desalination difficult. Large-scale desalination has also increased carbon dioxide emissions in the GCC, contributing to one of the world's highest carbon emissions per capita. In addition, desalinated seawater has some practical problems, including the difficulty and cost of storage in large strategic reserves (Strategy& 2014).

The process of desalination significantly impacts the environment, including damage to marine environments due to the release of brine and other chemicals into the sea and air pollution due to high emissions of CO_2 and other harmful gases (Almansouri and de Châtel 2016).

7.6.2 Price of Water

Mahmoud and Abdallah (2013) reported that the total cost for one cubic meter of desalinized water is about SR 5.5–6.6 for delivery to the consumers. The price of water for business is based on usage. The government's aim is to reduce the subsidies on water to offset the declining revenues of the Kingdom due to drop in the oil prices. The government plans to bring behavior change in civil society and consumers by making them realize how costly this natural resource is and how important it is to conserve resources. The current price of water and wastewater for industrial uses is SAR 6.0 [US$1.6] per cubic meter of water.

Tariffs set for urban water in the Kingdom do not encourage consumers to use water wisely. It is realized that low tariffs could be a disincentive for its conservation. Ouda (2013a, b) states that the KSA water tariff system was heavily subsidized by the government, as citizens were paying less than 5% of the water service cost. Due to the low cost recovery and the rapidly increasing demand, water subsidies pose a very severe stress on the country's revenues. After carefully studying the possible outcomes and impacts on consumers, the Ministry of Water and Electricity has adopted new tariff plans (with an increase of 50% for water and sanitation services provided for government bodies and large industrial and commercial establishments) to increase prices and improve sanitation services in all regions of the Kingdom. The ultimate objective of the plan was to reduce the high water consumption rates and rationalize its use.

The Saudi government has taken action to actively reduce local water use with a 5-year plan to gradually increase water tariffs (Almansouri and de Châtel 2016). The block tariff has been chosen to save the small and medium household consumers from the price hikes. The block rate system is viewed (in principal) as an efficient water demand management tool if it is well structured. The Minister of Environment, Water and Agriculture stated that the present tariffs do not cover more than 30–35% of the actual cost of the water pumped to households (Saudi Gazette 2016a, b). On the other hand, an economic research think-tank at the Gulf Research Center believes that since April 2016, water prices have climbed as much as 500% for Saudi nationals (Arab News 2015; Arabian Business 2016). Currently, the water tariffs in Saudi Arabia are one of the lowest in the world, ranging from US\$ 0.06 to US\$ 0.10 per m^3. However, the water prices are expected to rise, as the demand for water increases in the near future.

7.6.3 Water Quality

Al-Omran et al. (2015) collected one hundred and eighty drinking water samples from the main water network as well as underground and upper household tanks from the five zones of Riyadh governorate including Riyadh. The water quality for drinking with respect to chemical characteristics was found to be acceptable. However, analyses revealed some microbial contamination. The study revealed that most of the water zones had excellent quality and were quite safe for drinking (class I) purposes. The remaining water zones were few, but their water was unsuitable for drinking (class V) due to microbial contamination. Another study was conducted by Alsawalha (2017) to determine the quality of domestic drinking water by collecting 40 water samples from the main water source of the public network in homes in Jubail city. The study revealed that water samples were safe for drinking purposes and were of high quality based on the standard specifications outlined by the World Health Organization (WHO). A survey was made to assess the quality of 4255 well samples collected from the seven regions of the Saudi Arabia. Water samples were analyzed for total dissolved salts, nitrate, nitrite, ammonium, and fecal coliforms. Results showed that human and animal wastes were the primary source of pollution present in the water samples obtained from wells surveyed (El-din et al. 1994).

7.6.4 Water Losses Through Leakage Due to Old and Broken Infrastructure

Saudi Arabia's National Water Council reported that leakage reduction activities have already resulted in significant cost savings. However, the Global Water Intelligence estimated that 30% of desalinated water does not actually reach end customers and is lost to the environment during distribution (Global Water Market 2017). This results not only in wasted water and lost revenue, but also in unnecessary energy losses due to the energy required to desalinate and pump water that does not make it to the customer. Due to the lack of procedures in place and low level of knowledge on quantifying, it remains difficult to make the real estimates on the overall volume of this "nonrevenue" water. In Riyadh, where there are meters, an estimated 34% of the water is lost. The losses are split into 21% physical losses and 13% commercial losses due to incorrect measurements and illegal connections. In addition, about 24% of water is not paid by the consumers (MEWA 2018).

7.6.5 Unaccounted for Water

It is estimated that 20% of distributed water is unaccounted for in Saudi Arabia, due to outdated and in some cases deteriorating infrastructure (US-Saudi Arabian Business Council 2009).

7.6.6 Climate Change

While Saudi Arabia is striving to improve food self-sufficiency, it faces serious challenges due to such factors as drought, a limited agricultural sector, scarce water resources, and the serious impact of climate change (AFED 2014). Climate change will have a heavy impact on water resources and farming systems in the future. A simple change in temperature can have a significant impact on agriculture. The Kingdom has experienced prolonged periods of drought and flooding due to climate change. Climate change affects the frequency and intensity of extreme weather events. The extreme precipitation differences with increasing floods affect the Arab region. Severe floods affected Saudi Arabia and Yemen between 2008 and 2009, with estimated economic losses of approximately $ 1300 million USD (EM-DAT, www.emdat.be). Drought and climate change can have a significant impact on agriculture and the countryside in Saudi Arabia, with serious implications for food security, rural–urban migration, and social stability (Darfaoui et al. 2010).

7.6.7 Low Irrigation Efficiency

In 2000, approximately 66% irrigation activities used modern irrigation technologies, while traditional surface irrigation represented 34%. Despite the high adoption rates of modern irrigation technologies, inefficient use of water still prevails and can be commonly noticed on farms (Lovelle 2015). The irrigation efficiency is 50% at the present time (National Water Strategy 2016).

7.6.8 *Water and the Ecosystems and the Brine Disposal*

Desalination is a water technology that is gaining increasing importance for addressing water needs, but it is costly and energy intensive and further strains the environment with brine disposal. The brine produced during the desalination process causes damages to the local sea environment where the brine is discharged. In order for desalination to be considered a sustainable water solution, both issues must be successfully addressed (Xevgenos et al. 2016). According to studies, up to 800,000 cubic meters of brine is discharged into the Red Sea every day from the city of Jeddah alone. Nutrients, chemicals, and other waste products are all contributing to higher levels of pollution in the sea adjacent to the city. Such discharge is damaging the coral reefs and reducing fish stock. The risk to the Red Sea ecosystem posed by wastewater from Jeddah and other major industrial centers will increase in the coming years as Saudi Arabia's manufacturing expansion program gains pace (Oxford Business Group 2013). In fact, in coastal desalination plants, water pollution is the main problem. Desalination plants cause water pollution by disposing of hot brine, which has both thermal and saline impacts on the seawater into which it is released. Desalination raises the water temperature by about 9 °C. These plants have a serious impact on marine life in the area of desalination and cause environmental damage.

7.7 The Kingdom's New National Water Strategy

With the active involvement of all the stakeholders and the brainstorming of the related ministries, water companies and the community representatives, the government of Saudi Arabia has been able to put together a strategy that defines and identifies the most practical, efficient ways to produce, provide, and regulate water (Alturki 2015). A long-term national water strategy also appreciates the long-term capital investments in water projects over the next decade (MEWA 2018). The water strategy launched by the Ministry of Environment, Water and Agriculture comprises the following salient features:

In order to know the exact number of wells and to regulate them, farmers would be required to have authorization, and obtain a license for digging of wells from the Ministry of Environment, Agriculture and Water. Efforts are also underway to develop appropriate mechanisms to determine the shares for various uses of water. The Ministry has devised and is promoting an integrated management system to adequately utilize water resources economically and meet water demand. The Ministry attaches great importance to the uplift and upgradation of its human resources and to equip them with scientific and technical skills. Therefore, serious efforts have been made to elevate their working abilities, capacities, and development. Measures have also been taken to make water and sanitation services reliable and to achieve the highest levels of efficiency. To keep the business in good standing, new water tariffs to incentivize rational use of water for all purposes and its conservation have been successfully implemented since 2016.

As public–private partnerships have been successful in many sectors and in many parts of the world, KSA has also offered incentives and simplified procedures to encourage the private sector to invest in partnership in the water sector. The Kingdom highly encourages using and having greater reliance on renewable energy sources, particularly solar energy, in the water sector and aims at setting higher goals in this respect.

Treated water is gradually gaining acceptance both by the public and by the Ministry. To realize this objective, specialized projects and water plants in the larger cities have been constructed. This wastewater is made available through a network of pipelines to farms in the villages and rural areas.

The Ministry has also adopted a strategy that aims to promote and diversify agricultural production. Farmers are encouraged to cultivate crops with low water requirements and employing water-saving technologies.

Although the Kingdom has adopted the rollback policy, farmers remain the prime focus as they help ensure food security. The Ministry is helping them to use advanced irrigation techniques such as drip and sprinkler systems to improve irrigation efficiency in order to reduce pressure, enhance water-use efficiency, and conserve water resources. The Ministry has mapped the major agricultural areas identifying the crops and available water resources. Based on these maps, farmers are required to shift some of the fodder and cereals areas to lower water requirement crops. The Ministry has also constituted the Water Extension Centre to educate farmers about more sustainable uses of water. Dedicated measures have been taken to intensify the methods of rationalizing water uses for all purposes. Therefore, the Ministry has made a strong commitment to achieve a fair balance between water development resources and water consumption measures and uses.

Most of the water withdrawn comes from fossil water held in the deep aquifers of the Kingdom. Some observers predict that such water resources may not last more than 25 years (Mahmoud and Abdula 2013). The agricultural sector remains the principal consumer of water resources in the Kingdom. The government wants to regulate water at the farm level by installing water meters to assess farmers according to the volume consumed in irrigating crops. Farmers in the past have enjoyed generous subsidies and interest-free loans. Selling water and billing for the water based on consumption rates would be controversial and the farmers may not readily accept the change.

The Ministry has adopted nationwide campaigns to educate consumers and to create public awareness on water conservation measures. Adopting a cost-recovery approach in the provision of urban water and treatment services to reduce consumption will lessen the financial and environmental burden on the Kingdom's budget, realize sustainable development, and ensure water and energy security (The State of the Environment 2017). The expansion of treated wastewater reuse should go through setting standards and adopting quality control measures to check its suitability and reliability for safe use. Finally, flash floods generate huge amounts of surface water. Any simple but effective rainwater harvesting (RWH) structure can help improve the sustainability of the water supply in arid and semi-arid regions (Almazroui et al. 2017).

7.8 Energy–Water Nexus

Energy–water nexus refers to the relationship and interaction between energy and water. Water is used to generate electricity and produce energy. In Saudi Arabia, water is mainly used to generate energy for the cooling of thermal power plants and for the extraction, transport, and processing of fuel. On the other hand, energy remains essential in producing fresh desalinated water. It is used to power systems that collect, transport, distribute, and treat water (World Energy Outlook 2014; The Water-Energy Nexus 2016). Xylem (2014) considers energy generation the second largest water consumer after agriculture in the KSA and such energy use is expected to increase in the next 15–20 years. Water is used in power plants for electricity production, especially for cooling. This has significant implications for global water supply and environmental health in surface water bodies. The most common cooling systems are cooling towers and single-stage cooling (Rambo et al. 2017).

7.9 Water Policy and Economic Considerations

According to the Saline Water Conversion Corporation (2014), the companies consume an equivalent of 80 million barrels of oil a year. This high consumption prompted and compelled the Saline Water Conversion Corporation to employ more energy-efficient technologies. In the past, the reverse osmosis and thermal technologies were using the 70:30 ratios. In order to economize on costs and reduce energy consumption, numerous plans for large reverse osmosis projects on both coasts have been undertaken. The projects of Jeddah and Ras al-Khair were first examples in this direction. SWCC plans to reduce the fuel consumption of desalination plants by 10% by 2018 (Al-monitor 2016). In addition, Saudi Arabia prefers the "combined cycle power plants" technology for desalination plants so they produce electricity and desalinated water.

With an increasing number of new desalination projects, the KSA is also looking into trying new technologies. In January 2015, Advanced Water Technology (AWT), a newly established company announced that it would develop the world's first large solar-powered desalination plant in association with a global technology company. AWT is also establishing ties with the City of Science and Technology of King Abdulaziz (KACST) and the King Abdullah University of Science and Technology. Also, a new desalination plant has been built to deliver 60,000 cubic meters per day to the city of Al Khafji in northeastern Saudi Arabia. The plant uses the RO technique and an associated 15 MW photovoltaic (PV) plant would be capable of meeting all the energy needs of the installation. The use of renewable energy will also reduce the operating costs of the plant. KACST predicts that this pilot project will pave the way for a future in which desalinated water generated by renewable energy will replace fossil-fueled desalination in the Kingdom. KACST would become the owner of both the desalination facility and the photovoltaic plant in Al Khafji upon their completion. Plans to build a solar-powered plant to produce 300,000 m^3/day on the Red Sea coast are underway (Oxford Business Group 2017).

7.10 Conclusions and Recommendations

High population growth with increased per capita water consumption poses a threat to the sustainable use of water resources. Generous subsidies on fuel and desalinated water are resulting in depletion of the energy resources, as 10–20% of the oil reserves of the kingdom are being consumed by desalination plants. Factors like low water prices and the increased leakage of the water supply network increase the cost of drinking water, increase the pressure on desalination plants and sewage treatment plants, as well as air pollution emissions from desalination plants (The State of the Environment 2017). A consulting firm (Strategy& 2014) recommends that the KSA achieve greater sustainability in its water sector by addressing three issues: excessive water demand, inadequate water supplies, and ineffective institutional frameworks. The water scarcity and its consumption by the society are sensitive issues. In order to generate more revenues, the government was taking consumer-friendly steps. The country needs a well-thought and planned water policy. However, a first step would be to create awareness and bring behavior change among youth.

The KSA is making concerted efforts but it needs to move faster and intensify its efforts. All the steps taken to realize the fair balance in demand and supply ensure the sustainability of meager water resources, raise water prices, and revisit the desalination water production by using renewable energy and by lowering production cost, reducing CO_2 emissions, and disposing wastewater safely without harming the marine ecosystem. Actions to achieve sustainability in the water sector must be implemented in an integrated manner because water has an impact on multiple industries, from energy to agriculture and recreation. The reality is that KSA is really in search of a more *"Sustainable Water Culture."*

Acknowledgments

1. The authors are thankful to Saudi Society of Agricultural Sciences for its praise-worthy support and the resultant write-up.
2. Sincere thanks are due to Prof. Dr. Michael R. Reed, Director – International Programs for Agriculture at the University of Kentucky, USA and Dr. R. Kirby Barrick, Emeritus Professor, Agricultural Education and Communication University of Florida, USA, for reviewing the initial drafts, making helpful comments and valuable suggestions.

References

Abderrahman, W. A. (2001). Energy and water in arid developing countries: Saudi Arabia, a case-study. *International Journal of Water Resources Development, 17*(2), 247–255.
Abderrahman, W. A. (2006). Groundwater resources management in Saudi Arabia. Special presentation on water conservation workshop, Al Khobar, Saudi Arabia.

Abderrahman, W. A., & Al-Harazin, I. M. (2008). Assessment of climate changes on water resources in the Kingdom of Saudi Arabia. In *Proceedings of the GCC environment and sustainable development symposium, Dhahran, Saudi Arabia*, 28–30 January 2008. pp. D-1-1–D-1-13.

Abderrahman, W., & Rasheeduddin, M. (2001, March). Management of groundwater resources in a coastal belt aquifer system of Saudi Arabia. *Water International, 26*(1), 40–50. https://doi.org/10.1080/02508060108686885

Addressing Water Scarcity Through Recycling and Reuse: A Menu for Policymakers. https://smartcitiescouncil.com/resources/addressing-water-scarcity-through-recycling-and-reuse-menu-policymaker

AFED. (2014). *Food security challenges and prospects Arab environment report 7*. Report of the Arab Forum for Environment and Development 2014. Edited by Abdul-Karim Sadik, Mahmoud EL-Solh and Najib Saab. Available at: http://www.afedonline.org/Report2014/E/Binder-eng.pdf

Al-Awsat. (2018). *Saudi Arabia: 18 Projects Worth $3.2 Bn to Enhance Quality of Drinking Water*. https://aawsat.com/english/home/article/1521621/saudi-arabia-18-projects-worth-32-bn-enhance-quality-drinking-water

Al-Hussayen, A. (2007). *Minister of Water & Electricity, "Water Situation in Saudi Arabia and MOWE's Initiatives Speech"*. Saudi Arabia Water Environment Association Workshop, 2007.

Al-Hussayen, A. (2009). *Inaugural speech to Saudi water & power forum*. Jeddah: Saudi Water & Power Forum.

Al-Ibrahim, A. M. (2013). Seawater desalination: The strategic choice for Saudi Arabia, Desalination. *Water Treat, 51*, 1–4.

Al Jassim, N., Ansari, M. I., Harb, M., & Hong, P. Y. (2015). Removal of bacterial contaminants and antibiotic resistance genes by conventional wastewater treatment processes in Saudi Arabia: Is the treated wastewater safe to reuse for agricultural irrigation? *Water Research, 73*. https://doi.org/10.1016/j.watres.2015.01.036.

Aljaloud, A. A. (2010, August). *Reuse of wastewater for irrigation in Saudi Arabia and its effect on soil and plant*. Paper presented at the 19th World congress of soil science, soil solutions for a changing world, Brisbane, Australia.

Almansouri, A., & de Châtel, F. (2016). *Saudi Arabia's search for a more sustainable water culture*. Saudi Arabia's great thirst. Water around the Mediterranean. Chapter 5, (pp. 45–51). Available at: http://www.revolve-water.com/saudi-arabia-water-resources/

Almazroui, M., Islam, M. N., Balkhair, K. S., Şen, Z., & Masood, A. (2017). Rainwater harvesting possibility under climate change: A basin-scale case study over western province of Saudi Arabia. *Atmospheric Research, 189*, 11–23.

Al-monitor. (2016). *What Saudi Arabia is doing to save water?* Available at: https://www.al-monitor.com/pulse/originals/2016/04/saudi-arabia-self-sufficiency-water-policy.html

Al-Omran, A., Al-Barakah, F., Altuquq, A., Aly, A., & Nadeem, M. (2015). Drinking water quality assessment and water quality index of Riyadh, Saudi Arabia. *Water Quality Research Journal, 50*(3), 287–296. https://doi.org/10.2166/wqrjc.2015.039.

Al-Salamah, I. S., Ghazaw, Y. M., & Ghumman, A. R. (2011). Groundwater modeling of Saq Aquifer in Buraydah, Al Qassim for better water management strategies. *Environmental Monitoring and Assessment, 173*, 851–860.

Al-Saud, M. B. I. (2013, February 4–6). N*ational water strategy: The roadmap for sustainability, efficiency, and security of water future [sic] in the K. S. A. Water Arabia 2013*. Link here, last accessed December 2013.

Alsawalha, M. (2017). Assessing drinking water quality in Jubail Industrial City, Saudi Arabia. *American Journal of Water Resources, 5*(5), 142–145. https://doi.org/10.12691/ajwr-5-5-1.

Al-Subaiee. (2018). *Water in Saudi Arabia*. Presentation made at the King Saud University on the world Water Day, Riyadh, Saudi Arabia.

Al-Suhaimy, U. (2013, July 2). *Saudi Arabia: the desalination nation*. Asharq Al-Awsat. Available at: http://www.aawsat.net/2013/07/article55308131. Accessed 17 Aug 2015

Al Zawad, F., & Aksakal, A. (2009). *Impacts of climate change on water resources in Saudi Arabia*. https://doi.org/10.1007/978-1-4419-1017-2_33

Alturki, F. K. (2015, June). *Promoting sustainable development through environmental law: Prospects for Saudi Arabia.* SJD dissertation, Pace University School of Law. http://digitalcommons.pace.edu/lawdissertations/17/

Al-Zahrani, K. H., & Baig, M. B. (2011). Water in the Kingdom of Saudi Arabia: Sustainable Management Options. *The Journal of Animal & Plant Sciences, 21,* 601–604.

Al-Zahrani, M., Chowdhury, S., & Abo-Monasar, A. (2015). Augmentation of surface water sources from spatially distributed rainfall in Saudi Arabia. *Journal of Water Reuse and Desalination, 5,* 391–406.

Arab News. (2015). *Water tariff to rise 50% next month in Saudi Arabia.* Published — Monday 2 November 2015. Accessed 19 Jan 2018. Available at: http://www.arabnews.com/saudi-arabia/news/829286

Arabian Business. (2014a, March 26). *Saudi plans $533 m of water infrastructure projects.* http://www.arabianbusiness.com/saudi-plans-533m-of-water-infrastructure-projects-543922.html

Arabian Business. (2014b). *Saudi commissions world's biggest desalination plant.* Appeared on 23 April, 2014. Available at: http://www.arabianbusiness.com/saudi-commissions-world-s-biggest-desalination-plant-547616.html

Arabian Business. (2016, May 17). *New Saudi minister to set 'affordable' water tariff.* New environment minister Abdulrahman Al Fadli draws up new tariff following complaints over high bills. http://www.arabianbusiness.com/new-saudi-minister-set-affordable-water-tariff-631981.html

Argaam. (2018). *Saudi Arabia's drinking water consumption rises 8% in 2018.* https://www.argaam.com/en/article/articledetail/id/1312492

Baig, M. B., & Straquadine, G. S. (2014). Chapter 7: Sustainable agriculture and rural development in the kingdom of Saudi Arabia: Implications for agricultural extension and education. In M. Behnassi, M. Syomiti, R. Gopichandran, & K. Shelat (Eds.), *Vulnerability of agriculture, water and fisheries to climate change: Toward sustainable adaptation strategies* (pp. 101–116). Dordrecht: Springer Science +Business Media B.V.

Central Statistics Agency Statistics. (2017). *Saudi Reports and Statistics - Saudi - National Portal.* https://www.saudi.gov.sa

Central Department of Statistics & Information (CDSI). (2010). *Population & housing census for 1431 a.h (2010 a.d) findings.* Riyadh, KSA. Retrieved January 10, 2015, from http://www.cdsi.gov.sa/

Chowdhury, S., & Al-Zahrani, M. (2013). Implications of climate change on water resources in Saudi Arabia. *Arabian Journal for Science and Engineering, 38*(8), 1959–1971.

Chowdhury, S., & Al-Zahrani, M. (2015). Characterizing water resources and trends of sector wise water consumptions in Saudi Arabia. *J King Saud University, Engineering Science, 27,* 68–82.

Darfaoui, C., El-Mostafa, A., & Al-Assiri, A. (2010). *Response to climate change in the Kingdom of Saudi Arabia.* A Report Prepared for FAO-RNE. Available online at: http://www.fao.org/forestry/29157-0d03d7abbb7f341972e8c6ebd2b25a181.pdf

DeNicola, E., Aburizaiza, O. S., Siddique, A., Khwaja, H., & Carpenter, D. O. (2015). Climate change and water scarcity: the case of Saudi Arabia. *Annals of Global Health, 81*(3), 342–353.

Drewes, J. E., Patricio Roa Garduno, C., & Amy, G. L. (2012). Water reuse in the kingdom of Saudi Arabiastatus, prospects and research needs. *Water Science & Technology-Water Supply, 12,* 926–936.

El-Din, M. N., Madany, I., Al-Tayaran, A., Al-Jubair, A. H., & Gomaa, A. (1994). Quality of Water from Some Wells in Saudi Arabia. *Water, Air, and Soil Pollution, 66,* 135–143.

Enssle, C., & Freedman, J. (2017). *Addressing water scarcity in Saudi Arabia: policy options for continued success.* www.suezwatertechnologies.com

FAO. (1997). *Irrigation in the near east region in figures.* Rome: FAO Country Report Saudi Arabia 9 C. FAO Land and Water Division, Food and Agriculture Organization of the United Nations.

Food and Agriculture Organization (FAO). (2009). *Saudi Arabia. Irrigation in the Middle East regions in figures.* Aquastat Survey -2008. FAO Land and Water Division Report 34, 325–337. Edited by Karen Freken.

General Authority for Statistics. (2016). *Percentage of water extracted or produced from fresh water by source (surface, underground, desalinated, treated) in Saudi Kingdom in 2010–2016.* Available at: https://www.stats.gov.sa/sites/default/files/percentage_of_water_extracted_or_produced_from_fresh_water_by_source_surface_underground_desalinated_treated_in_saudi_kingdom_in_2010-2016en.pdf

General Statistics Authority (GSA). (2018). *Environmental indicators. Statistics on water resources of Saudi Arabia.* Available at: https://www.stats.gov.sa/en/node/10131

GIEWS. (2018). *Country brief – Saudi Arabia.* Available at: http://www.fao.org/giews/country-brief/country/SAU/pdf/SAU.pdf. Accessed 12 Mar 2019.

Global Water Market. (2017). Meeting the world's water and wastewater needs until 2020. Volume 4: Middle East and Africa. Saudi Arabia, (pp. 1379–1385).

IMF. (2015, June). *Is the glass half empty or half full? Issues in managing water challenges and policy instruments.* IMF staff discussion note. Kalpana Kochhar, Catherine Pattillo, Yan Sun, Nujin Suphaphiphat, Andrew Swiston, Robert Tchaidze, Benedict Clements, Stefania Fabrizio, Valentina Flamini, Laure Redifer, Harald Finger, and an IMF Staff Team.

Kajenthira, A., Siddiq, A., & Anadon, L. D. (2012). A new case for promoting wastewater reuse in Saudi Arabia: bringing energy into the water eq. *Journal of Environmental Management, 102*, 184–192.

KAPSARC. (2016). Policy options for reducing water for agriculture in Saudi Arabia. In C. Napoli, B. Wise, D. Wogan, & L. Yaseen (Eds.), The King Abdullah Petroleum Studies and Research Center (KAPSARC) (pp. 1–27). Riyadh, Saudi Arabia. March 2016 / KS-1630-DP024A.

Khrais, R. (2016, April 4). What Saudi Arabia is doing to save water. *Al-Monitor of the Middle East.* Available at: https://www.al-monitor.com/pulse/originals/2016/04/saudi-arabia-self-sufficiency-water-policy.html

KAUST (King Abdullah University of Science & Technology). (2011). *KAUST industry and collaboration program (kicp), the kicp annual strategy study: Promoting wastewater reclamation and reuse in the kingdom of Saudi Arabia: Technology trend, innovation needs, and business opportunities.* Jeddah: KAUST.

Lippman, T. W. (2014, May 24). Biggest Mideast crisis: Growing water scarcity. *Business Mirror.* Available at: http://www.ipsnews.net/2014/05/ biggest-mideast-crisis-probably-dontknow-enough/. Accessed 17 Aug 2015.

Lovelle, M. (2015). *Food and Water Security in the Kingdom of Saudi Arabia.* Future Directions International.

Mahmoud, M. S. A., & Abdalla, S. M. A. (2013). Management of Infrastructure for Water and Petroleum Demand in KSA By GIS. *Innovative Systems Design and Engineering, 4*(14), 49–75.

MEP. (2010). *The ninth development plan 2010–2014. Ministry of economy and planning documents.* Riyadh: KSA Government.

MEWA (Ministry of Environment, Water and Agriculture). (2017). *Annual report.* Riyadh, Saudi Arabia, 2017. Available online: https://www.mewa.gov.sa/en/Pages/default.aspx. Accessed on 27 Dec 2017

Ministry of Environment, Water and Agriculture (MEWA). (2018). *National water strategy.* Available at: https://mewa.gov.sa

Ministry of Water and Electricity. (2012) Supporting documents for King Hassan II Great Water Prize.

MWE: Ministry of Water and Electricity Annual Report. Riyadh, Saudi Arabia. Available at: http://www.mowe.gov.sa/ENIndex.aspx. Accessed 22 June 2015.

Nachet, S., & Aoun, M.-C. (2015). *The Saudi electricity sector: Pressing issues and challenges.* The Institut français des relations internationales (Ifri).

Nachet, S., Aoun, M. C., & Saudi electricity sector. (2015). *The Saudi electricity sector: Pressing issues and challenges 30 March 2015 Centre Énergie Note de l'Ifri. one.* ISBN: 978-2-36567-370-9 Website: ifri.org. Brussels Belgium and Paris France. Available at: https://www.ifri.org/sites/default/files/atoms/files/note_arabie_saoudite_vf.pdf

Napoli, C., Wise, B., Wogan, D., & Yaseen, L. (2016). *Policy option for reducing water for agriculture in Saudi Arabia.* March 2016/KS-1630-DP024A.

National Water Company. (2008). *Management contract for water supply and wastewater collection services in the city of Jeddah, Kingdom of Saudi Arabia*. Riyadh: National Water Company.

National Water Company. (2012). *National water company – passion for excellence*. Riyadh: National Water Company.

National Water Company. (2015). *National Water Company website*. Retrieved January 10, 2015, from http://www.nwc.com.sa/English/Pages/default.aspx.

National Water Strategy. (2016). *Ministry of Environment, Water and Agriculture*. https://mewa.gov.sa/en/Ministry/Deputy%20Ministries/TheWaterAgency/Topics/Pages/Strategy.aspx

NWC. (2017). *Treated Sewage Effluent (TSE)*. https://www.nwc.com.sa/English/OurOperations/Business-Development-Initiatives/Pages/Treated-Sewage-Effluent-(TSE).aspx

NWC. (2018). National Water Company. *Treated waste effluent*. Available at: https://www.nwc.com.sa/English/OurOperations/Business-Development-Initiatives/Pages/Treated-Sewage-Effluent-(TSE).aspx.

Ouda, O. K. (2013a). Towards assessment of Saudi Arabia public awareness of water shortage problem. *Resources and Environment, 3*(1), 10–13.

Ouda, O. K. (2013b). Review of Saudi Arabia Municipal Water Tariff. *World Environment, 3*(2), 66–70. https://doi.org/10.5923/j.env.20130302.05. http://www.pmu.edu.sa/kcfinder/upload/files/Review_of_Saudi_Arabia_Municipal_Water_Tariff.pdf.

Ouda, O. K., Shawesh, A., Al-Olabi, T., Younes, F., & Al-Waked, R. (2013). Review of domestic water conservation practices in Saudi Arabia. *Applied Water Science, 3*(4), 689–699.

Ouda, O. K. (2014a). Impacts of agricultural policy on irrigation water demand: A case study of Saudi Arabia. *International Journal of Water Resources Development, 30*(2), 282–292.

Ouda, O. K. (2014b). Water demand versus supply in Saudi Arabia: Current and future challenges. *International Journal of Water Resources Development, 30*(2), 335–344.

Oxford Business Group. (2013, May 27). Saudi Arabia: Water demands continue to rise. *Saudi Arabia Economic News*. https://oxfordbusinessgroup.com/news/saudi-arabia-water-demands-continue-rise

Oxford Business Group. (2017). *Saudi Arabia expands its desalination capacity*. https://oxfordbusinessgroup.com/analysis/world-leader-efforts-under-way-expand-desalination-capacity

Rambo, K. A., Warsinger, D. M., Shanbhogue, S. J., & Ghoniem, A. F. (2017). Water-energy Nexus in Saudi Arabia. *Energy Procedia, 105*, 3837–3843.

Saline Water Conversion Corporation (SWCC). (2010). *Annual report for operation & maintenance*. Saline Water Conversion Corporation, (pp. 20–35).

Saline Water Conversion Corporation (SWCC). (2012). *Annual report for operation & maintenance*. Riyadh: Saline Water Conversion Corporation.

Saline Water Conversion Corporation. (2014). *Annual report*. Riyadh: Saline Water Conversion Corporation. https://www.swcc.gov.sa/english/MediaCenter/SWCCPublications/PUBLICATION%20FILES/ANNUAL%20REPORT%202014ENC02348BB-A1D8-4029-BFE3-136FA70DAB0B.PDF

"Saudi Arabia Water Report Q4 2013." (2013, August). Business Monitor International.

Saudi Gazette. (2016a). *New water tariff to be acceptable to all: Fadli*. http://saudigazette.com.sa/article/155166/New-water-tariff-to-be-acceptable-to-all-Fadli

Saudi Gazette. (2016b). *Innovative approach to water management a 'must' in KSA*. http://saudigazette.com.sa/article/146334/Innovative-approach-to-water-management-a-must-in-KSA.

Saudi Geological Survey. (2012). *Kingdom of Saudi Arabia numbers and facts. Saudi Geographical Survey* (1st ed.). Riyadh: KSA Government.

Scotney, T. (2014) Global Water Market 2014 (Saudi Arabia). Global Water Intelligence (GWI), Media Analytics Ltd., April 2013. Also see: KAUST, 2010-11, cited below.

Staff writer. Water Resources. Royal embassy of Saudi Arabia, Washington, DC. Available at: https://www.saudiembassy.net/about/country-information/agriculture_water/Water_Resources.aspx. Accessed 17 Aug 2015.

Stratfor. (2017). *Where Saudi Arabia's Water Disappears To*. https://worldview.stratfor.com/article/where-saudi-arabias-water-disappears

7 Water Resources in the Kingdom of Saudi Arabia: Challenges and Strategies... 159

Tarawneh, Q. Y., & Chowdhury, S. (2018). Trends of climate change in Saudi Arabia: implications on water resources. *Climate, 6*(1), 8. https://doi.org/10.3390/cli6010008

Tchobanoglous, G. (1995). Decentralized systems for wastewater management. Decentralized systems for wastewater management. In: *24th annual WEAO technical symposium*, Toronto, Canada.

The State of the Environment. (2017). Responsibilities and achievements.

The Water-Energy Nexus: Challenges and Opportunities | Department of Energy. [Online]. Available: http://www.energy.gov/downloads/water-energy-nexus-challenges-and-opportunities. Accessed 13 Feb 2016.

Toze, S. (2004). Reuse of effluent water benefits and risks. "New directions for a diverse planet." In: *Proceedings of the 4th international crop science congress*, September 26 e October 1, 2004. Brisbane, Australia.

US -Saudi Arabian Business Council. (2009). *The water sector in the Kingdom of Saudi Arabia.* Vienna: U.S.-Saudi Arabian Business council.

World Bank. (2004). *Kingdom of Saudi Arabia, assessment of the current water resource management situation. Phase I* (Vol. 1). World Bank, Washington, DC.

World Bank. (2005, March 31). *A water sector assessment report on Countries of the cooperation council of the Arab State of the Gulf.* Report No. 32539-MNA.

World Bank - Global Economy Report. (2018). *Global economic prospects.* Chapter 2. http://pubdocs.worldbank.org/en/233201557323177690/Global-Economic-Prospects-June-2019-Analysis-MENA.pdf

World Energy Outlook. (2014). OECD/IEA.

Xevgenos, D., Moustakas, K., Malamis, D., & Loizidou, M. (2016). An overview on desalination & sustainability: renewable energy-driven desalination and brine management. *Desalination and Water Treatment, 57*(2016), 2304–2314. Available at: www.deswater.com January. https://doi.org/10.1080/19443994.2014.984927.

Xylem. (2014). *Water use in oil and gas: Trends in oil and gas production globally.* Editors of Xylem.

Dr. Mirza Barjees Baig is Professor of Agricultural Extension and Rural Development at the King Saud University, Riyadh, Saudi Arabia. He has received his education in both social and natural sciences from the USA. He earned his MS degree in International Agricultural Extension in 1992 from the Utah State University, Logan, Utah, USA, and was placed on the "Roll of Honor." He completed his PhD in Extension Education for Natural Resource Management from the University of Idaho, Moscow, Idaho, USA. During his doctoral program, he was honored with "1995 Outstanding Graduate Student Award." He has published extensively in the national and international journals. He has also presented extension education and natural resource management extensively at various international conferences. Particularly, issues like degradation of natural resources, deteriorating environment, and their relationship with society/community are his areas of interest. He has attempted to develop strategies for conserving natural resources, promoting environment, and developing sustainable communities through rural development programs. He started his scientific career in 1983 as a Researcher at the Pakistan Agricultural Research Council, Islamabad, Pakistan. He has been associated with the University of Guelph, Ontario, Canada, as the Special Graduate Faculty in the School of Environmental Design and Rural Planning from 2000 to 2005. He served as a Foreign Professor at the Allama Iqbal Open University (AIOU) through Higher Education Commission of Pakistan, from 2005 to 2009. He is the member of the IUCN Commission on Environmental, Economic, and Social Policy (CEESP). He is also the Member of the Assessment Committee of the Intergovernmental Education Organization, United Nations, EDU Administrative Office, Brussels, Belgium, and of many national and international professional organizations. Moreover, he serves on the editorial boards of many international journals.

Dr. Yahya Alotibi is an Assistant Professor at the Department of Agricultural Extension, King Saud University, Riyadh, Saudi Arabia. Recently, he has earned his PhD from the Iowa State University, USA. His dissertation (Perceived Distance and Instructional Design in Online Agriculture and Life Science Courses) received a Research Excellence Award from the Iowa State University (ISU), USA. Before entering into the doctoral program at the Iowa State University, he earned Graduate Certificate in Instructional Technology/Instructional Design, Learning Environments, and Educational Program Planning. He served as an Instructional Designer at the Iowa State University and helped the Department of Kinesiology to convert one of their courses from a traditional face-to-face course to hybrid course, using blended learning strategies. He also completed his Master of Education (Med) program at the Clemson University in 2014. His professional interests include climate smart agriculture, sustainable water management under stressed environments, food waste, innovations in agricultural extension education, communication, program evaluation, strategic planning, and educational technology.

Prof. Dr. Gary S. Straquadine serves as the Interim Chancellor, Vice-Chancellor (Academic Programs), and Vice Provost at the Utah State University Eastern, Price, UT USA. He completed his PhD from the Ohio State University, USA. Presently, he also leads the applied sciences division of the USU Eastern campus. He is responsible for faculty development and evaluation, program enhancement, and accreditation. In addition to his heavy administrative assignments, he manages to find time to teach some undergraduate and graduate courses and supervise graduate student research.

Being an Extension Educator, he has a passion for the economic development of the community through education and has also successfully developed significant relations with agricultural leadership in the private and public sector. He has also served as the Chair, Agricultural Communication, Education, and Leadership, at the Ohio State University, USA. Before accepting the present position as the Vice Provost, he served on many positions as the Department Head, Associate Dean, Dean and Executive Director, USU Tooele Regional Camp, and Vice Provost (Academic). His professional interests include extension education, sustainable agriculture, food security, statistics in education, community development, motivation of youth, and outreach educational programs. He has also helped several underdeveloping countries improving their agriculture and educational programs. He has a very strong passion for the healthy ecosystems for the healthy communities, higher education, and international development programs. He has been honored with numerous awards and honor for making significant contributions to the society and science. His research has been published as the book chapters in the prestigious books and scientific articles in the high impact journals.

Dr. Abed Alataway is an Assistant Professor at the Prince Sultan Institute for Environmental, Water and Desert Research, King Saud University, Saudi Arabia. He earned his PhD in Water and Environmental Management from Newcastle University, UK; his MSc in Water and Environmental Management from Loughborough University, UK; and his bachelor's degree in Agricultural Engineering from King Saud University, Saudi Arabia.

Chapter 8
Water Policy in Tunisia

Mohamed Salah Bachta and Jamel Ben Nasr

Abstract In Tunisia, climate varies from Mediterranean to arid and semi-arid. Water resources are characterized by scarcity and a pronounced irregularity. The water supply policy, with nearly 40% of agricultural investment in the 1980s, allowed to develop an irrigated agriculture and to create an undeniable productive potential. Despite the performances in irrigated agriculture, the water sector still suffers from economic, social, ecological, and institutional problems. The scarcity of water induced by the increasing demands of various economic sectors and climate change effects leads to the questioning of current allocations of the water resources. Given the economic model adopted and the current governance, water scarcity would be a major threat for sustainability.

Keywords Water demand · Scarcity · Climate change · Sustainability · Governance

8.1 Introduction

Tunisia has a predominantly semi-arid climate. Two-thirds of the country is semi-arid to arid. Both the climate and especially precipitation levels fluctuate greatly throughout the year and from year to year. These extreme variations observed can be seen in the form of droughts or floods, which can have serious consequences both ecologically and economically, including soil impoverishment, overexploitation of aquifers in the case of prolonged droughts, as well as variations in agricultural production and therefore, farmers' income.

M. S. Bachta · J. Ben Nasr (✉)
Department on Agricultural Economics and Management and Agrifood, National Agronomic Institute of Tunisia, Carthage University, Tunis, Tunisia

© Springer Nature Switzerland AG 2020
S. Zekri (ed.), *Water Policies in MENA Countries*, Global Issues in Water Policy 23,
https://doi.org/10.1007/978-3-030-29274-4_8

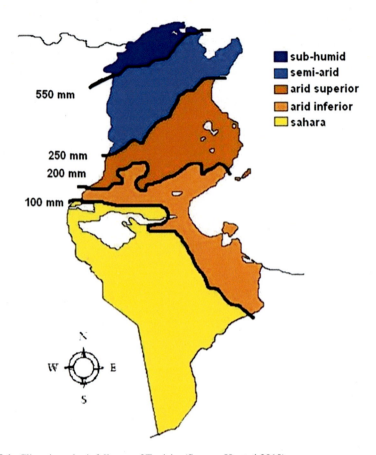

Fig. 8.1 Climatic and rainfall map of Tunisia. (Source: Hentati 2010)

To mitigate these fluctuations, the Tunisian government developed a policy aiming to increase supply of water resources and create irrigated areas. The water resources are estimated at 4.874 billion m³/year from which 2.7 billion m³/year are surface water. The rate per capita, 385 m³/capita/year, shows that Tunisia has been in a situation of absolute water scarcity since the 1990s. Moreover, the availability of water, particularly surface water, has many regional disparities. Whereas in the north, which makes up around 25% of the total area of the country, rain varies from 400 to 1500 mm, approximately 60% of the country receives less than 200 mm a year (Fig. 8.1).

These variations necessitated the creation of a large infrastructure of surface water reservoirs in order to transfer water to areas with water scarcity, in particular, the coastline.

By the end of 2014, infrastructure included 33 dams with a total cumulative storage capacity of 2.237 Km³, 253 hill dams with a total capacity of 266 million m³,

8 Water Policy in Tunisia

and 902 hilly lakes with a total capacity of 93 million m^3. This hydraulic infrastructure was developed during the pre-revolution (before 2011) sociopolitical context, with a strong State and a submissive society. Three Master Plans were developed and led to an inventory of water resources and the identification of uses. The legislation in the form of the water code of 1975 authorized the State to manage the water resources (assignment, transfer, and pricing). An omnipresent Administration was in charge of managing the infrastructure in place. This Administration exercised a discretionary power; consequently, all decisions were imposed via a top-down mechanism. In this natural and sociopolitical context, irrigated agriculture was designed and implemented. This technical and sociopolitical arrangement, however, suffers from many failings and seems to have reached its limit.

8.2 Legal and Institutional Aspects of the Water Sector

The analysis of the Tunisian water policies shows that the institutional framework of water management has undergone significant transformations. These transformations can be summed up in three decisive periods: The period of colonization (1881–1956) during which we note the deconstruction of the ancestral system of collective management and the establishment of a new system reflecting the despotism of the colonial authorities. The second period extends from the year of independence (1956) till the late 1980s (1989), and the nationalization of water resources through the creation of the development boards and the promulgation of the water code are the main features of this period. The third period runs from 1989 until the 2011 revolution, starting the water management transfer to the stakeholders in 1989 and the genesis of a series of emergence of new organizations and dismantling of the public ones. This development illustrates the progressive disengagement of the State in favor of a "collective" management of water resources.

The establishment by the colonial authority was marked by a process of centralization of natural resources, particularly water resources. An operation of destruction and destabilization of the traditional system of governance has begun to gradually install traditional management methods marked by the expropriation of local populations in favor of a despotic colonial administration. A despotic colonial system gradually replaced local institutions. New instruments have been created and imposed. The first decree sealing the centralization of water resources was promulgated in 1885. Legal forms of irrigation water management institutions have increased ranging from trade union associations of owners of oases created between 1912 and 1920 in southern Tunisia to special associations of hydraulic interest instituted in 1933 whose powers are similar to those of AIC which appeared in 1936. We also note the unification of regulatory texts. Nevertheless, these imposed institutional transformations clashed with disengagement and almost total rejection of the local population.

In the aftermath of independence, the Tunisian State relied on the establishment of development offices as an instrument for the management of the irrigated areas and supervision of farmers. The first office, the office development of the Medjerda Valley, (OMV) was established in 1958. It is in the mid-1970s that the water policy was most clearly defined with strong objectives for resource mobilization and allocation. This period was marked by the promulgation of the Water Code in 1975 (Law No. 75-16 of March 31, 1975) by the development and implementation of master plans and by the establishment of irrigation development boards in the main regions of the country. The promulgated Water Code clearly and definitively established the ownership of all water resources, transforming old property rights into use rights. Resource allocation priorities are set by the State through the Master Plans aimed at the precise identification of available resources and to plan their mobilization and allocation among the different uses in the three main regions of the country: the north, central, and south.

Since 1975, the management of the public hydraulic domain (DPH) was put under the auspices of the Ministry of Agriculture. Several technical departments were established, the most important of which are (i) the Directorate General of Water Resources (DGRE), which ensures the day-to-day management; (ii) the General Directorate of Dams and Large Hydraulic Operations (DGBGTH) for the feasibility and construction of dams, the exploitation of dams, and the development of large irrigated areas; (iii) the General Direction of the Rural Engineering and the management of water resources (DGGREE) for the implementation of the policy of development of irrigated areas, drinking water supply in rural areas, and the promotion of the water saving and associative management of water systems. This system is complemented by the Office of Hydraulic Planning attached to the Minister's cabinet, in charge of inventories of water resources and uses, as well as the planning and scheduling of water allocations (Fig. 8.2).

The Development Offices are regional establishments whose mission is the management of hydraulic equipment and the supervision of irrigators. External demands by international donors have forced the Political System (SP) to change the approach to undertake water management through market mechanisms or collective actions at the level of irrigator communities. The development offices in charge of irrigated areas were dissolved and user associations (AIC) were created from scratch. Thus, since 1990, management has been entrusted to "AIC", which were later transformed into "GIC" collective interest groups and then into "GDAP" Agricultural Development and Fishing Groups (Law 99-43 of May 14, 1999 as well as Law N ° 2004-24 of 15.03.2004) (Al Atiri 2006).

Wanting to keep control of water resources, the state has triply dominated the associations created (financially, technically, and politically). In these circumstances, the sustainability of these associations in terms of financial, political, and social issues seems problematic. The consequences of this institutional mechanism on resources are obvious: overexploitation and poor maintenance of equipment (Bachta et al. 2001).

The analysis of these institutional transformations and their impact on the performance of the management and exploitation of water resources has been the subject of several research projects conducted in Tunisia. The first approach considers the institutional mechanisms as major determinants of the action of the actors and giv-

8 Water Policy in Tunisia

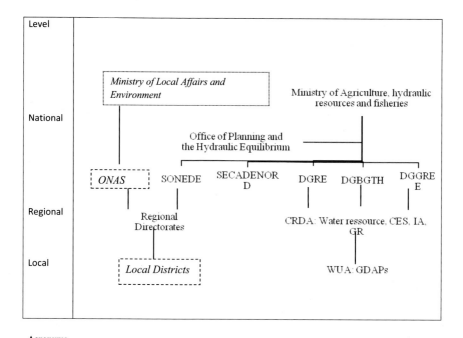

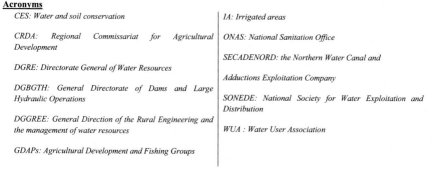

Fig. 8.2 Main actors of water management in Tunisia. (Source: Self-elaboration)

ing the absolute power to the organizational structure. The performance of the managers is considered as an endogenous variable. According to this methodological approach, Mokrani (2012) explains that the water allocation rule and the number of irrigators have proved to have an important explanatory power of the recovery rates recorded by the groups (GDAP) investigated. The second approach admits that the actors have the latitude to influence the structures in place and therefore enjoy a certain freedom in their choices (Ben Nasr 2015; Rekik and Bachta 2016). Ben Nasr and Bachta (2016) have shown that a greater involvement of farmers in the selection of the members of the management committees of these GDAPs and their participation on the definition of rules to adopt are likely to improve the performance of the GDAPs.

8.3 Agricultural Water

In Tunisia, two major types of farms are observed, the rain-fed and the irrigated ones. Under various incentives (economic, financial, and institutional) developed within the framework of agricultural policy, these two types of farming have achieved great advances. The irrigated area went from 62,000 ha in the 1960s to 450,000 ha in 2014. Rain-fed farms, in particular those in the central and southern regions, also expanded considerably, particularly the plantation of olive and almond trees. It should be noted that this type of farming is totally dependent on climatic conditions, drought in particular.

The undertaken investments resulted in an undeniable economic growth. The average annual growth rate of the agricultural gross domestic product was estimated at 3.5%. The irrigated farming system contributed 40% to the agricultural GDP on average and the value of agricultural production has seen significant interannual fluctuations (Hamdane and Bachta 2015). Despite the development of the other sectors of the economy, agriculture still represents a relatively significant part of the GDP with 11% on average. It is important to note that this performance required heavy investments in water reservoirs which are predominantly used for agricultural purposes.

8.3.1 Mobilization of Water Resources

A strategy of water supply was implemented and great efforts of surface waters storage marked the decades of 1970s and 1980s with the construction of large dams. During the following decade (1990s), small and medium-sized reservoirs represented the main public water investments. The objective was to reach control of 95% of surface water and groundwater resources. Table 8.1 shows that such a rate has been achieved for the surface water resources. Currently, Tunisia has 33 dams with a total capacity of 2.242 billion m^3, 253 small dams, and 837 hill lakes. Nearly half of the water resources are stored in large dams, 5% comes from small dams and mountain lakes, and the rest is from aquifers (ITES 2014).

In Tunisia, surface water is distributed among the three main regions of the country, as follows: Ninety percent is available in the north with 2190 million m^3 per year of which 1796 million m^3 is of high quality with salinity lower than 1.5 g/l. The center is with 320 million m^3 including 153 million m^3 with a salinity less than 1.5 g/l. The southern part of the country is with 120 million m^3 per year including 5 million m^3 with a salinity of less than 1.5 g/l.

Groundwater resources represent nearly 44% of all of the total water resources. The deep aquifers have the largest storage capacity. Their exploitation is reserved to the public authorities. Private entities can reach it only after obtaining proper authorization. However, illicit drillings abstracting these resources can be seen. On the other hand, no authorization is required to access aquifers, and thus, they are much

8 Water Policy in Tunisia

Table 8.1 Changes in surface water captured in Millions m³

	Potential available water resources	Actual reservoirs capacity				
		1990	2000	2005	2010	2015
Large dams	2170	1170	1688	1927	2080	2170
Hill dams	195	5	125	160	190	195
Hilly Lakes	135	7	38	62	88	94
Total	2500	1182	1851	2149	2358	2459
Rate of surface water captured		44	69	80	87	93

Source DGRE (2015)

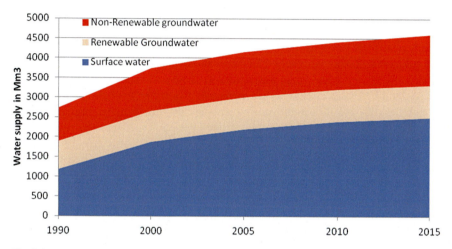

Fig. 8.3 Water resources in Mm³. (DGRE 2015)

demanded. However, illegal drillings that capture these resources are observable. These practices are reported in the central (Kairouan) and southern (Kebili) regions – regions where water resources are the most in demand. In the region of Kairouan, the farmers manage to circumvent the prohibition texts by proceeding with deepening of their surface wells. In Kebili, local societies still have a certain capacity for organization, which gives them opportunities to collectively carry out locally determined actions such as extensions of palm plantations.

In parallel with the increase in the volumes of water withdrawn, the number of wells has steadily increased, reaching in 2010 more than 130,000 surface wells and 9300 deep wells (Hamdane 2014).

Figure 8.3 shows the evolution of the mobilization of all the exploited water resources.

Some 30 deep aquifers are overabstracted with a rate of abstraction 1.68 and a volume estimated at 315 Mm³ (World Bank, cited by Elloumi 2016). For the shallow aquifers, the volume abstracted was estimated in 2005 to be 548 Mm³ while the renewable volume is about 355 Mm³. Thus, the deficit is estimate at 193 Mm³"

(Hamdane, cited by Elloumi 2016). The overabstraction resulted in the degradation of the quality of the water, in particular an increase of salinity, as well as on the drawdown of the level of water table, pushing the farmers to deepen their well and complicating the sustainable management of the resource. As a response, the response from the government was a supply side policy by starting a strategy to recharge some of the aquifers with volumes varying between 30 and 70 Mm^3 per year.

8.3.2 Reuse of Treated Wastewater

The nonconventional water resources are limited mostly to treated wastewater and are estimated at approximately 232 Mm^3 in 2012 representing 5% of the total resources. It is estimated that wastewater in 2030 will reach 300 Mm^3 per year (ITES 2011). Sewage discharges are increasingly bulky and numerous, resulting in a more complex pollution to the receiving bodies, making them increasingly vulnerable. This pollution seriously threatens groundwater and surface water resources as well as beaches and other water environments.

As shown in Fig. 8.4 below, wastewater has not been the source of such a development.

From this figure, it can be seen that the agricultural use of treated wastewater remained, overall, below forecasts. The quality of the treatments and the constraints imposed by the law limiting the crop mix to be practiced are the variables put forward by the farmers to explain this underutilization. The reuse of the wastewater is more significant in the regions of the south and the center (Sfax and Mahdia) where

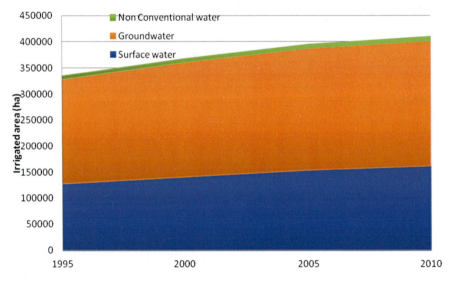

Fig. 8.4 Evolution of the irrigated areas by irrigation sources. (ITES 2011)

the cultural choices of farmers, olive trees, and forage crops correspond to the crops allowed by the legislation in place. To these variables should be added the absence of other alternative water resources which are otherwise completely absent, at least rare and coveted by other uses. Improving the quality of these waters through further processing, in accordance with established and institutionalized standards, could encourage farmers to use them. Thus, improved bacteriological quality of the treated water reduces the health risk for farmers and encourages them to use these waters (El Amami et al. 2003). The protection of the environment and public health could provide additional legitimacy for such improvements.

8.3.3 Pricing

The pricing of irrigation water has undergone successive adaptations. In fact, at the beginning of the service of the main irrigated areas, the tariffs were kept at low levels to encourage farmers to adopt irrigation. The cost of providing water services for irrigation has weighed heavily on the state budget. The agricultural structural adjustment program and the devolution of the irrigation institutions to farmers led to the search for water cost recovery rates as an important objective in the management of the irrigated areas. Successive increases to the irrigation water prices have been decided. A rigorous policy of annual increase of irrigation water prices has been implemented since 1991 and prices have been increased by 9% in real terms. Such rates are generally higher than those recorded by the prices of other factors of production and those of the producer prices of most agricultural crops (Hamdane and Bachta 2015). This policy has been observed in most governorates, but in a less sustained manner in some central and southern governorates.

As a replacement for the governmental irrigation (Offices de Development), user associations were created and then generalized. During the last decade, each association has been authorized to set its selling price level and its water pricing method, taking into account its budget balance. Currently, there are different modes of pricing in the irrigated areas. Each tariff system is influenced by the type of management of the irrigated area, the equipment in place, and the cropping systems. In the irrigated areas equipped with meters, a single pricing method has been adopted. In the absence of meters, the pricing becomes lump sum. Billing takes into account only the areas declared and the price per cubic meter. It is obvious that this alternative pricing does not encourage the farmers to value the resource and does not reduce the waste. Irrigation water pricing can also be per ha. It is practiced in southern oases where water demands are relatively uniform. This mode covers 7% of the total Tunisian irrigated areas. Sales by the hour is another mode of pricing practiced in the old irrigated areas with open networks of the lower valley of Medjerda river and the irrigated areas of the south and the center.

Finally, the two blocks pricing mode consists of a fixed term payable per ha and a variable term depending on the actual consumption of water. Dual pricing is an incentive to increase the demand for water. Indeed, the average price of m^3 will decrease when the quantity consumed increases. It is therefore recommended in irrigated areas where intensification rates remain low, especially in winter and where the supply of water is higher than the demand. In addition, this pricing method provides fixed revenues to the manager and is a cost-sharing instrument within an area among all owners. Thus, in order to guarantee regular revenue to the managing institution of the irrigated areas, the move to block pricing was introduced by public authorities from 1999 to 2000 on the large irrigated areas of the north in particular.

This diversity of pricing methods leads to significant variations in water prices paid by irrigators. Zekri et al. (1996) have shown that groundwater prices vary considerably between regions and in the same region. The authors explain these differences in water prices by the effects of economies of scale.

The policy of increasing water tariffs faced several constraints including the low rate of intensification, the abandonment of irrigation in winter, and the shortage of water. In addition, these tariffs are set without any link with the payment capacities of farmers. Such a demand management tool is totally ignored by irrigated areas managers. Kefi (1999) and Kefi et al. (2004) showed that in the irrigated areas of Kairouan, farmers' ability to pay depends mainly on the crops grown, the irrigation technology adopted, and the management method. They concluded that the increase in tariffs must be accompanied by measures that improve the economic valuation of water: more efficient irrigation practices and mainly the introduction of value-added crops.

8.3.4 Irrigation Efficiency

Studies conducted in different irrigated areas in Tunisia still show relatively low technical and economic efficiency scores (Albouchi et al. 2003; Belloumi and Matoussi 2006; Dhehibi et al. 2007; Dhebi and Telleria 2012; Ben Nasr et al. 2016). Thus, efficiency gains are still achievable through the implementation of production technologies that save water, increase land-use rates, improve crop yields, the limitation of land fragmentation, and the reduction of interfarm conflicts. The different components of the increase in efficiency should lead to a better valuation of this resource increasingly coveted by other sectors of the economy.

In this logic of improving the value of irrigation water, the State has implemented panoply of reforms and actions, including, in particular, (i) reforms of demand management through rehabilitation and the modernization of the hydraulic infrastructure; (ii) the financial and technical incentive to save water, within the framework of a National Program for Water Economy (PNEE); and (iii) the creation and institutional strengthening to empower users in the management of water distribution and the maintenance of tertiary networks.

The National Program for Water Economy in Irrigation (PNEE) was adopted in 1995 This program has been widely disseminated with the extension of the modernization of irrigation water systems, which has been favored by the political decision to increase the grant allowed to modern equipment of irrigation [40, 50, and 60% of the cost of investments, respectively, for large, medium, and small farms]. From 1995 to 2010, the total investments made under this program were estimated at 937 MD, including 468 MD in financial incentives from the State. A relative progress has been recorded, thanks to the program under consideration; the rate of equipment in water-saving technology has increased from around 37% of the total area irrigated in 1995 to 83% in 2010 and 87% in 2012. The total area equipped in 2012 by the PNEE is estimated at 365000 ha of which 90,000 ha is in improved gravity irrigation, 116,000 in sprinkling, and 160,000 in localized. Localized and sprinkler irrigation has introduced innovations such as fertigation.

It is hard to believe that this program has resulted in a reduction in the overall demand for irrigation water. Consumptions per hectare have certainly decreased. The adoption of this program led, on the one hand, to an extension of irrigated surfaces and, on the other hand, an increased energy required by new irrigation techniques.

Most of the groundwater comes from deep aquifers in the south, the largest of which is nonrenewable fossil groundwater (610 million m^3/year, which represents 42% of deep aquifer resources). The total contribution of these aquifers is estimated at 1429 million m^3 per year.

Groundwater quality is generally poor. Nearly 84% of these resources have salinity levels in excess of 1.5 g/l. The overabstraction of these aquifers, estimated on average at around 103%, is one of the determinants of the deterioration of their quality. This average hides important interregional variations. The governorates of Nabeul with 154%, Kairouan with 123%, Kasserine with 112%, and Kebili with 179% are the most affected.

The major consequences of this overabstraction are a significant lowering of the level and prospects of disappearance of shallow aquifers and the gradual deterioration of the chemical quality of water. These consequences would result in excessive increases in treatment costs of these waters and an aggravation of the deterioration of their quality. Such developments would make future uses of these waters truly problematic and could put an end to their use.

Groundwater, whose potential is estimated at 846 Mm^3, is suffering from overexploitation, which has reached 139% for some groundwater in the center of the country. All pumped volume are used for irrigation, as only 2% is less than 1.5 g/l.

For the future, the resource-use adequacy considered by the "Eau XXI" study (see Table 8.2) focuses on the rather moderate evolution of domestic, industrial, and tourist water demands. Volumes allocated to the irrigated sector are revised downward due to competition from other sectors of use. Measures to encourage farmers to use water-saving techniques and to adopt lower-water demanding crop varieties, rear-season crops, and short-cycle varieties would lead to a significant reduction in the allocation of irrigation water. Thus, the average allocation per hectare would decrease from around 6000 m^3/ha in the 1990s to 4350 m^3 by 2030. The total allocation to the sector would thus be revised from 2150 Mm^3 in the 1990s to 2035 Mm^3 at the same horizon. The only possible compensation for the irrigated sector is the use of treated wastewater; a volume of around 220 Mm^3 would be made available

Table 8.2 Evolution of water demand by use (Mm^3: Million cubic meters)

Water Uses	Horizons				
	2010	2015	2020	2025	2030
1. Urban water					
Domestic	381	410	438	464	491
Industrial	136	150	164	183	203
Tourism sector	31	33	36	39	41
Subtotal	**548**	**593**	**638**	**686**	**735**
2. Agricultural water	**2140**	**2115**	**2082**	**2058**	**2035**
Irrigated areas (1000 ha)	410	417	433	450	467
Volume in m^3/ha	5200	5000	4800	4600	4350
Total	**2688**	**2708**	**2720**	**2744**	**2770**

Source: MARHP (2000)

Table 8.3 Evolution of the share of irrigated agriculture in exports

	Value of exports (1000 DT)				
Year	1995	2000	2005	2011	2015
Total agricultural exports	462,000	628,200	1,225,600	2,126,300	3,647,000
Irrigated agriculture					
Dates	58,300	52,700	131,500	297,500	445,300
Citrus	10,600	9800	15,200	18,700	23,000
Drift of fruits and vegetables	39,800	71,200	63,000	116,000	92,900
Vegetables	0	0	22,200	103,400	97,000
Total irrigated agriculture	*108,700*	*133,700*	*231,900*	*535,600*	*658,200*
Share of irrigated agriculture (%)	*23,53*	*21,28*	*18,92*	*25,19*	*18,05*

Source (ONAGRI, several years)DT: Tunisian dinar

for irrigation by 2030. Such use could concern land not far from wastewater treatment plants, usually located outside urban centers.

The content of Table 8.3 will be interpreted as the result of projections aimed at testing hypotheses of changes in the various components of the water resources-use balance sheet. These simulations should test the feasibility of a balance between supply and demand for water.

These projections have highlighted the necessity to reduce the need for water per irrigated ha to observe the water balance. It is important to question the technical and sociopolitical feasibility of such an assumption, the implementation of which seems to be taking into account the controlled technologies and the capacity of irrigators to manage situations of shortage, which is particularly problematic. In fact, the limited availability of surface water recorded during year 2017 led irrigation water managers to ration the supply of this resource.

8.4 Water Property Rights

The water code of 1975 converted all private property rights of water into use rights as a consequence of instituting the nationalization of the water resources which is an eminently political decision. Through this decision, the state has given itself the legitimacy to reconfigure the interests involved. The water code indeed authorized the State to control all the resources, their allocation between sectors and between regions and this, without reference to society. For example, water transects from the north to the center have been decided and implemented and the supply of domestic water is considered a priority over other uses.

8.5 Food Security vs Virtual Water/Food Imports

The flagship products exported by Tunisia are, with the exception of olive oil, particularly demanding in water. We talk about dates and citrus fruits. While palm trees are irrigated by fossil water, citrus fruits require the transfer of water from the north via the Medjerda-Cap Bon canal. The selection of current production mixes has allowed irrigated agriculture to make a significant contribution to total exports of agricultural products. Table 8.3 traces the evolution of this contribution.

This table shows an increase in the share of irrigated crops in total agricultural exports. Despite this steady growth, the contribution of the irrigated sector remains below that of olive oil estimated at around 40% on average of agricultural exports. Cereals with low irrigation requirements constitute the bulk of our imports of agricultural products. The food security interpreted by the government as the trade balance of agricultural and food products could result in a negative balance of the exchange in terms of virtual water.

8.6 Water Salinity

In Tunisia, conventional water resources are naturally affected to varying degrees of salinity. About 74% of the country's surface water has a salinity less than 1.5 g/l (82% of the waters in the north, 43% of the waters in the center, and 38% of the waters in the south). With regard to groundwater, only 8% of the resources have a salinity lower than 1.5 g/l and 21% have 4 g/l and more. About 20% of deep underground resources have a salinity less than 1.5 g/l, 57% have a salinity between 1.5 and 3 g/l, and the remaining 23% have a salinity exceeding 3 g/l. In summary, the salinity of waters exploited for irrigation purposes is as follows (Table 8.4).

Table 8.4 Salinity classes of conventional resources used in irrigation

Class of salinity	Volume in Mm^3	% of volume
S < 1.5 g/l	527	25%
1.5 < S < 3 g/l	791	37%
3 < S < 5 g/l	586	28%
S > 5 g/l	220	10%
Total	2124	100%

8.7 Urban and Drinking Water

Urban water with its drinkable and industrial components is considered a priority use compared to agricultural uses. It must also be of better quality. Under the effects of demographic growth, especially urbanized populations, but also the improvement of the standard of living of the population and diversification of the Tunisian economy, the share of this water in total uses will only increase in the future. This increase will be to the detriment of the agricultural sector (MARHP 2000).

The rate of access to drinking water is relatively satisfactory in Tunisia. In fact, the drinking water supply rate in urban areas is 100%. These national averages hide inter- and intraregional disparities. In rural areas, it rose from 30% in 1985 to 93.4% in 2012. Despite this remarkable increase and improvement, the rural service rate remains particularly low in the northern regions of the country (Bizerte, Beja, Kef, Jendouba and Siliana), although 80% of the surface water comes from these regions (BPEH 2013). Another form of regional inequity problem is confirmed here. These governorates have the lowest service rates (87%). Both the dispersal of habitats and the isolation due to the topography in rural areas of these regions do not facilitate the supply of the conquered populations.

The unconventional water resources considered are those derived from seawater desalination. Desalination of brackish water has been already taken into account in groundwater uses. Three seawater desalination plants were considered: that of Djerba with a capacity of 18.5 Mm^3/year that will be allocated to drinking water and that started operation in late 2018; a desalination plant in Sfax with a capacity of 73 Mm^3/year of which only 50% will be available in 2020 and the rest by 2030; and that of Zarrat, governorate of Gabes, with a capacity of 18.25 Mm^3 which will be totally dedicated to drinking water and available from 2020. A seawater desalination plant for the industrial needs of the Tunisian Chemical Group in Gabes is scheduled for 2020 with a capacity of 10.95 Mm^3/year.

Domestic water is metered and billed according to a scale of tiered pricing with several quarterly water consumption bands. The pricing system, which is the same for all the country, has seven consumption bands with a single tariff per bracket. Rates range from 0.2 dinar/m^3 for the first social block (0 to 20 m^3/quarter) to 1.315 dinar/m^3 for consumption greater than 500 m^3/quarter. It should be noted that a single tariff of 1315 Tunisian Dinar/m^3 is applicable to the tourism sector (SONEDE 2017).

8.8 Water and the Environment

ONAS (National Sanitation Office) is the only operator responsible for the collection and treatment of wastewater. Around 90% of the urban population is connected to sanitation networks. The number of inhabitants connected to the sewerage network is estimated at 6.3 million in 170 communes supported by ONAS. With a total of 110 wastewater treatment units, the total volume treated is 232 Mm^3/year in 2012 (BPEH 2013). Given the potential for rain-fed agriculture, treated wastewater is little used in the northern regions. In some cases, they are therefore surplus and a source of pollution. In the southern and central regions, the use of these waters is limited by the degree of treatment below the required standards; handling also remains a source of pollution.

The environmental water component accounts for about 2% of total available water. This water is essential for the maintenance of the ecosystems and hence for sustainable development. It should not suffer from reduction and should even be increased as much as possible (ITES 2011).

Regarding the allocation of soil irrigated by the saline water, there are about 60,000 ha of land sensitive to salinization and 75,000 ha subject to an enhancement of the aquifers in the irrigated areas in north and south in particular. On the other hand, it is estimated that about 60% of the public irrigated land in Tunisia is moderately to highly sensitive to secondary salinization following irrigation; this rate reaches 86% in the private irrigated areas. These salinity risk categories are shown in Fig. 8.5 below.

In addition to these uses with socioeconomic objectives, it is important to highlight the needs of eco-systems in blue and green water. Blue water is used in some cases as a source of environmental balance. Lake Ichkeul illustrates the case of an ecosystem dependent on a complex hydrological balance. This balance results from seasonal freshwater inputs from six streams flowing into the lake and water movements between the lake and the sea via the Tinja Canal.

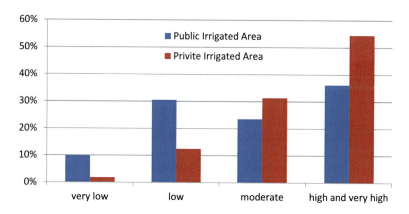

Fig. 8.5 Soil salinization risk classes. (MARHP- DG/ACTA 2005)

When the inflow of water is high, due to winter precipitation from October to March, the lake level increases, its salinity drops, and the excess water flows into the sea On the other hand, with low flows of freshwater, the level of water in the lake decreases, and the flow between this lake and the sea is reversed. The salinity of the waters of this lake will only increase. From 1992 to 2002, the hydrological balance was disrupted by two prolonged droughts and the diversion of a large volume of water from the streams that feed the lake. The lake's ecosystems were significantly affected, and there was a decline in lake productivity, a loss of key habitats, and a dramatic decrease in waterfowl populations using the lake and surrounding areas (ANPE 2004).

The intensification of agriculture can lead to the degradation of fertility and the pollution of irrigated land. A survey carried out in 2007 (Directorate of Soils, DG/ACTA) as part of a study on the evaluation of land degradation forms in irrigated areas revealed in particular the following facts relating to fertilization and pollution in these irrigated areas.

The decline in fertility of irrigated soils constitutes a dominant form of degradation in relation with the modification of certain physical, mineralogical, chemical, or biological properties of the soil (degradation of the structure due to lack of organic matter, loss of major nutrients, etc.). This change affects the overall quality of the soil and can lead to an often significant decrease in crop yields with significant economic consequences in the long term. The phenomenon in question affects about half of the public large irrigated areas surveyed.

Chemical pollution in irrigated areas is mainly due to the misuse or inadequate use of mineral fertilizers and crop protection products. It threatens in the long term 28% of the irrigated areas with localized low-to-medium pollution risks.

In some cases, particularly in the oases of south, hydromorphic signs due to lack of drainage systems are observable, such as the oases of Hezoua and Ibn Shabbat in Tozeur and the oases of Tarfaya in Kibili. Water logging damages soil structures and affects the growth and productivity of date palms. During the rainy years, in the northern region, particularly in the upper Madjerda Valley, 4000 to 5000 ha are submerged by runoff water (the case of Bou Salem in Jendouba). The water logging negatively affects crops, soil structure, as well as livestock and habitats.

It should be noted that the intensive use of nonrenewable resources threatens the sustainability of these resources. This situation essentially characterizes the oases and some of the irrigated areas which consume about a quarter of the water supply for irrigation in the country. Despite the implementation of careful planning and monitoring of these resources, signs of overuse and loss of the shallow aquifers are worrying for the future (see the case study of large illegal extensions of the oases at Kebili). This can seriously affect the salinity of the resource and increase the energy costs for abstraction. These risks are very serious for the south and constitute a challenge to the sustainability of date production, which is currently the mainstay of the economic development of certain regions of the south.

8.9 Energy–Water Nexus

The direct consumption of energy in the agricultural sector is estimated at about 7% of the country's total energy consumption, of which 2% is for irrigation. The consumption of this form of energy is generally fluctuating according to the climatic conditions. In fact, climatic factors determine the total area requirements of irrigated land and subsequently the energy requirements for irrigation. However, consumption of this form of energy has increased at a very low rate over the last decade, with an average of 1% per year. It is noted that total electricity consumption of the agricultural sector has been growing regularly in recent years, with an average of 6% per year. This is due to the replacement of diesel pumps with electric pumps. This increase in electricity consumption is due to the rapid growth of electrification for pumping due in part to the adoption of water-saving techniques. These technologies need an important quantity of energy (STEG Several years). The following graph relating to the number electrical agricultural connections to the electricity grid for Medium Voltage (MV) and Low Voltage (LV) shows such an evolution (Fig. 8.6).

Increasing use of energy, especially electricity, has not translated into real savings in water.

The desalinated water comes from brackish water or seawater. Thanks to the competitive cost of reverse osmosis desalination, it is becoming a valuable and economical source of water supply. Thus, the installed desalination capacity is currently estimated at about 138,000 m³/d.

Energy consumption by the agricultural sector and the water sector is continuously increasing. Between 2015 and 2016, there was an 11% increase in the consumption of energy for pumping fresh water and treated wastewater and a 6.7% increase for agricultural water pumping. The total electricity consumption for the agricultural sector reached 1286 Gwh in 2016 including 602 Gwh for agricultural activities and 684 Gwh for water pumping energy (STEG 2016).

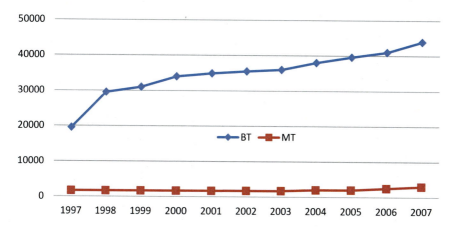

Fig. 8.6 Evolution of the number of subscribers to the electricity utility: Low voltage (LV) and medium voltage (MV). (Source Several years)

8.10 Special Issues

8.10.1 International Water: Conflicts, Negotiations, and Agreements

The two main aquifer systems in the south of Tunisia are transboundary resources with the majority of the corresponding surfaces located in Algeria (68%) and Libya (24.3%). Algeria withdraws 1.3 billion m^3 per year, or 60% of total withdrawals, followed by Tunisia with 0.540 billion (24.6%), followed finally by Libya with only 0.340 billion m^3 (15.4%).

The scenarios developed in the framework of the Northern Sahara Aquifer System project (SASS) all converge toward the more or less controlled increase in abstraction with significant effects in terms of the depletion of aquifers, up to 300 m for some aquifers. Total abstractions estimated at 600 Mm^3 per year in 1970 increased to 2.2 billion in 2000 and are expected to reach 8 billion m^3 in 2030 (OSS 2008, cited by Elloumi 2016). More broadly, the SASS program draws attention to the problem of these shared aquifers which is quite complex and which requires coordination between the authorities of the three countries.

In fact, the Tunisian authorities are caught between, on the one hand, a regulation of the abstractions by a strict control of the wells development and a race in the neighboring countries in order to make the most of such fossil and low renewable resources.

Oued Madjerda, which is a major source of surface water in Tunisia, is also shared with Algeria. Consultation mechanisms, particularly in case of flooding, are observed.

8.10.2 Climate Change Effects

Like all African countries, Tunisia will not escape the impacts of climate change phenomena despite its insignificant emissions of GHG compared to the global average. All the studies carried out over the last 10 years were based on the climate projections from the 2007 study on the adaptation strategy of the agricultural sector and ecosystems to climate change, which were also used in the second national communication published in 2014. These projections provided for an annual average increase in temperature concerning the whole country of +1.1 °C by 2030 and + 2.1 °C by 2050. The magnitude of this increase in temperature would vary from one area to another. It averages 1.6 °C in the northern region, 2.1 °C in the center, and 2.7 °C in the southern region.

Decrease in the annual volume of rainfall, which ranges from 10% in the north to 30% in the south by 2050, compared to the current situation. This decline will most likely be accompanied by an increase in the frequency and intensity of the extreme dry years.

These results are reinforced by the climatic projections published in 2015 by the National Institute of Meteorology (MNI), which, based on a dynamic downscaling of the results of the 4th IPCC report and on a mesh of 25 km², offer to Tunisia the first climate projections from one of the general circulation models validated for the case of Tunisia on 9 models.

The selected socioeconomic scenario describes a future world in which economic growth will be very rapid, the world population will reach a maximum in the middle of the century and then decline, and new, more efficient technologies will be introduced rapidly. It is a median scenario. The reference climatology is that of 1961–1990, which constitutes the historical period of reference of the climate. These are the climate projections that are used to assess the vulnerability of water resources to climate change.

This situation will worsen in the coming years as demand for water increases and as a result of climate change, with a decline in conventional water resources estimated at around 28% by 2030. The increase in extreme events (floods and droughts) will further weaken the country's water situation. These results are from the first initiative to assess the needs of agriculture and ecosystems for adaptation to climate change following the severe drought experienced by Tunisia between 1998 and 2002 (MARHP and GIZ 2007). In addition, following the accelerated sea-level rise, the salinization of the coastal aquifers would result in losses of about 50% of the current resources by 2030, that is, losses of nearly 152 Mm³. All studies carried out on the coastline confirm the fragility of coastal water resources and the major risks of seawater intrusion.

A recent study in 2016 assessed Tunisia's vulnerability to climate change. Vulnerability is approximated by the product of exposure, the ratio of water demands to water availability, and the ratio of sensitivity to adaptive capacity. Estimated at the level of the governorate, the vulnerability scores are all greater than unity, which means a proven vulnerability. Some governorates such as Kairouan, Sidi Bouzid, and Kebili are hot spots.

As part of this study, a prospective exercise was conducted. A basic scenario known as regional rebalancing has been adopted. This scenario simulates a process of population adjustment and unemployment rates. According to this simulated process in situations with and without CC, the populations of the western governorates will evolve at the national average rate, and the rates currently observed are lower than this average. The various water demand components estimated for this scenario were for the 2030 and 2050 horizons. As shown in Tables 8.5 and 8.6, total water demand will increase in the case of climate change. On the other hand, the decrease in demand for water by 2050 expressed by the agricultural sector would result from the fall in agricultural activity in favor of other sectors (industry and services), but also from the technological improvements linked to irrigation efficiency. With regard to water availability, Table 8.5 below summarizes blue water resources by category, taking into account the increase in aridity due to climate change.

We note that the water consumption of the agricultural sector will decline despite the effects of climate change. The predominant effect would be the diversification of economic activity which would result in an increase in the opportunity cost of the

Table 8.5 Supply of and demand for blue water at different horizons with CC (Mm³)

	Horizons	2014	2030 with CC	2050 with CC
Demand	Urban water	701	1121	1323
	Irrigation	2650	2296	2093
	Total demand for blue water	**3351**	**3417**	**3416**
Supply	Available surface water (Mm3)	1072	1176	843
	Shallow aquifers	746	720	705
	Deep aquifers	1429	1404	1380
	Total supply of blue water	**3247**	**3300**	**2928**

CC: Climate change; Mm³: Million cubic meters
Source: MARHP (2000)

Table 8.6 Supply of and demand for green water at different horizons – with Climate change scenarios (in Millions m³)

	Horizons	2014	2030	2050
Green water demand	Rain-fed agriculture	5069	4517	4150
	Forest and grass-land	9011	9034	9541
	Total green water demand	**14,080**	**13,551**	**13,691**
Green water supply	Rain-fed agriculture	11,033	10,880	10,770
	Forest and grass-land	9011	8800	8822
	Total green water supply	**20,044**	**19,680**	**19,592**

Source: MARHP (2000)
Mm: millions m³

labor force. This evolution will be to the detriment of agricultural activity. Such an economic context could imply the abandonment of certain irrigated areas or certain crops that have become uncompetitive with regard to the use of the primary factors of production, water, and labor. Citrus fruits and dates seem to be the most endangered.

For green water, supplies for rain-fed agriculture, rangelands, and forests are presented in Table 8.6 above.

The "green waters" – water indirectly available in the form of agricultural, animal, and vegetable production (rain-fed agriculture, rangeland, and forest) – total about 23 billion m³/year. There is therefore a huge reservoir that needs to be better exploited and mobilized to ensure and reinforce water security (ITES 2011).

While it is generally estimated that about 60% of rainwater is absorbed and evaporates and 39% flows and/or goes underground, in Tunisia, 90% evaporates and only 10% flows. The importance of green water is emphasized and there is a lack of research and interest in it (Besbes 2011).

The drought during the last years has led to overexploitation of the aquifers. The Ministry of Agriculture, Water Resources and Fisheries has programmed at the end of 2018 a measure of groundwater recharge. This recharge aims to improve the piezometric level and reduce the groundwater salinity. This recharge is possible by the releases from surface reservoirs, infiltration basins, and injections into surface wells. This recharge program targets a total of 48 million m³ and covers eight regions (see Table 8.7).

Table 8.7 Groundwater recharge in Tunisia in 2018

Regions		Quantity (millions m³)
North	Nabeul	10
	Ben Arous	2
	Bizerte	2
	Zaghouan	2
Center	ElFekka	10
	Kasserine	2
	Kairouan	10
South	Gafsa	10
Total		48

Source MARHP (2019)

8.11 Research

According to the summary report of the Strategic Study, Water 2050, a profound review of Agricultural Research and Development may be necessary in terms of defining missions, objectives, organization, work methods, and productivity. It is proposed to include research activities, with specific objectives, within the framework of the National Priority Programs. The technical and technologies proficiency in all areas of water particularly in the artificial recharge of aquifers, the tertiary treated wastewater, the silting of dams, the desalination of water, the selection of plants and seeds adapted to the Tunisian climate constitute the essential components guaranteeing national water security. Scientific research is to be developed in all these areas and it is necessary to rethink this effect (ITES 2011). In addition to these technical components, particular attention should be given to institutional aspects, governance, and water management, particularly at the local level.

8.12 Conclusions

The public authorities have chosen to invest in irrigated agriculture since it reduces the dependency on rainfall variability. They have made important investments in the supply of water and creation of irrigated areas. The relations of dominance between the state and society have given it considerable discretion. The agricultural activities created allowed undeniable performances. However, this agriculture suffers from threats of unsustainability. In fact, irrigated agriculture is threatened in its environmental, institutional, financial, and political components.

The water and soil resources are overexploited, especially groundwater, as well as risks of salinization and hydromorphic soils. These risks are more observable in areas with a long tradition of irrigation: the Cap Bon region, oases, etc.

In addition to these threats of the physical environment supporting the activities of irrigated agriculture, a three-tiered dualism characterizes farms. The first is of a structural nature, opposing small farms with little equipment to large production units with a balanced production structure. The second reflects the observable differences in the production techniques adopted. While the smaller farms use "traditional" technical itineraries, the other farms are in direct contact with technological advances on a global scale. The third component of this dualism is institutional in nature and reflects the differences in the relations of production. Salaried employment is the most widespread social relation at the level of large farms. The small ones practice various forms of sharecropping. This dualism is reflected in differences in the value of irrigation water. Small farms with an unbalanced structure ensure valuations far below the prices of water paid. As a result, there is a risk of nonpayment of water bills and, as a result, financial difficulties are often faced by water user associations.

On the political level, the revolution interpreted as a collective refusal of public choices has aggravated the questioning of the choices decided previously. There have been almost universal refusals to pay water prices. Protests have begun to rise against some interregional water transfers.

Finally, the guidelines adopted in terms of economic development, in particular, are being confronted by the new relations that the state maintains with the postrevolution society and, on the other hand, by the tensions over the water resources exacerbated by the effects of climate change on existing agriculture.

References

Al Atiri, R. (2006). Evolution institutionnelle et réglementaire de la gestion de l'eau en Tunisie. Vers une participation accrue des usagers de l'eau. L'avenir de l'agriculture irriguée en Méditerranée. Nouveaux arrangements institutionnels pour une gestion de la demande en eau, Cahors, France.

Albouchi, L., Bachta, M. S., & Le Grusse, P. (2003). Pour une meilleure valorisation globale de l'eau d'irrigation: une alternative de réallocation de la ressource sur des bases économiques: cas du bassin du Merguellil en Tunisie centrale [en ligne]. Séminaire PCSI [Programme de Recherches Coordonnées sur les Systèmes Irrigués] sur la Gestion Intégrée de l'Eau au Sein d'un Bassin Versant; 2003/12/02; Montpellier.

ANPE (Agence nationale de protection de l'environnement). (2004). Le suivi scientifique au parc national de l'Ichkeul. Année hydrologique 2002–2003. Tunisie.

Bachta, M. S., Le Goulven, P., Le Grusse, P., & Luc, J.-P. (2001). Environnement institutionnel et relations physiques pour une gestion intégrée de l'eau dans le milieu semi-aride méditerranéen. Le cas tunisien. In: Séminaire international Montpellier 2000, « Hydrologie des régions méditerranéennes », Montpellier, 11–13 octobre 2000. Paris, Unesco, PHI-V/Documents techniques en hydrologie, n° 51, pp. 177–186.

Belloumi, M., & Matoussi, M. S. (2006). A stochastic frontier approach for measuring technical efficiencies of date farms in Southern Tunisia. *Agricultural and Resource Economics Review, 35*(2), 285–298.

Ben Nasr, J. (2015). Gouvernance et performance de la gestion de l'eau d'irrigation en Tunisie: cas des périmètres irrigués de Nadhour-Zaghouan. Agronomie. Institut National Agronomique de Tunisie. Thèse conduite à l'INAT-Tunisie sous la direction de Pr. Bachta MS.

Ben Nasr, J., & Bachta, M. S. (2016). Gouvernance et performance de l'exploitation de l'eau dans les périmètres irrigués de Nadhour: Quels effets des rapports de force? Revue des régions arides.

Ben Nasr, J., Akkari, T., Fouzai, A., & Bachta, M. S. (2016). Le mode d'accès à l'eau d'irrigation un déterminant de l'efficacité des exploitations agricoles: Cas du périmètre irrigue de Sidi Ali Ben Salem, Kairouan-Tunisie. *Journal of New Sciences, Agriculture and Biotechnology, 29*(5), 1676–1686.

Besbes, M. (2011). L'eau en Tunisie: gestion de la rareté conjecture de pénurie. Conférence du 28 Mai 2011 à l'Auditorium de l'INAT. Tunis.

BPEH (Bureau de planification et des équilibres hydrauliques). (2013). L'alimentation en eau potable et l'assainissement en Tunisie. Note a élaborée en Novembre 2013 par BPEH (Ministère de l'agriculture des ressources hydrauliques et de la pêche). Available on this website: https://www.ohchr.org/Documents/Issues/Water/Handbook/Tunisia.pdf

DGRE (2015). Direction Générale des Ressources en Eau. Ressources en Eau en Tunisie. Rapports annuels

Dhebi, B., & Telleria, R. (2012). Irrigation water use efficiency and farm size in Tunisian agriculture: A parametric frontier analysis approach. *American Eurasian Journal of Agriculture and Environmental Science, 12*(10), 1364–1376.

Dhehibi, B., et al. (2007, September). Measuring irrigation water use efficiency using stochastic production frontier: An application on citrus producing farms in Tunisia. *African Journal of Agricultural and Resource Economics* 1(2), 1–15

El Amami, H., Natsoulis, D., & Xanthoulis, D. (2003). Evaluation économique du traitement des eaux usées traitées par épuvalisation. Proceedings des actes du séminaire International Réutilisation des eaux usées traitées et des sous produits: Optimisation, Valorisation & Durabilité (Tunis 24–25 septembre 2003).

Elloumi, M. (2016). *Groundwater governance in the Arab world*. IWMI project report N0.7.

Hamdane, A . (2014). La gestion des ressources en eau souterraines (nappes et aquifères) comme biens communs : Cas de la Tunisie, SCET, Tunisie, Version provisoire citée par Elloumi 2016.

Hamdane, A., & Bachta, M. S. (2015). Intensification de l'agriculture irriguée en Tunisie. FAO-Banque Mondiale. Programme Eau, Climat et Développement pour l'Afrique "Carte de vulnérabilité des ressources en eau et de leurs usages aux impacts du changement climatique en Tunisie »- Novembvre 2016.

Hentati, A. (2010). *Geomorphological study on erosion vulnerability of small hillside catchments in the semi-arid region of Tunisia*. PhD thesis in civil and environmental engineering, with specialization in hydrology. Tokyo Metropolitan University.

ITES. (2011). Institut Tunisien des Etudes Stratégiques. Etude Stratégique: Eau 2050 en Tunisie. Rapport de synthèse.

ITES. (2014). Etude stratégique : Système hydraulique de la Tunisie à l'horizon 2030, janvier 2014. p. 222.

Kefi, M. (1999) – Estimation de la rente économique de l'eau dans les PPI de Chébika, gouvernorat de Kairouan. Mémoire de PFE en Economie Rurale et Gestion sous la direction de Pr BACHTA MS.à l'INAT.

Kefi, M., Faysse, N., Le Goulven, P., & Bachta, M. S. (2004). Comportement des irrigants face à des changements d'accès à l'eau dans les périmètres irrigués de la plaine de Kairouan. In: Le Goulven P., Bouarfa S., Kuper M., 2004. Gestion intégrée de l'eau au sein d'un bassin versant. Actes de l'atelier du PCSI, 2–3 décembre 2003, Montpellier, France.

MARHP (Ministère de l'agriculture des ressources hydrauliques et de la pêche). (2019). Gestion pluriannuelle des épisodes de crue et de sécheresse dans le nord la Tunisie par référence aux années (2016-17-18).

MARHP, & Giz. (2007). Stratégie nationale d'adaptation du secteur agricole et des écosystèmes au changement climatique.

MARHP(Ministère de l'agriculture des ressources hydrauliques et de la pêche). (2000). EAU XXI-Stratégie à long terme du secteur de l'eau en Tunisie 2030, DGGR.

Mokrani, A. (2012). *Essai d'explication de la performance des associations d'irrigants : Gouvernorat de Nabeul*. Mémoire de mastère. Institut National Agronomique de Tunisie.

Rekik Bouguecha, S. & Bachta, M. S. (2016). Une dynamique co-évolutive « institution-innovation» et gains de productivité dans la gestion des ressources communes. Revue des Régions Arides n°40 (2/2016) – Numéro spécial : Actes des travaux du colloque international LOTH 2016 « Gouvernance et communication territoriales » Mahdia (Tunisie) 7, 8, 9 avril 2016

SONEDE. Société Nationale d'Exploitation et de Distribution des Eaux. (2017). Tarification de l'eau potable. Available on http://www.sonede.com.tn/index.php?id=111.

STEG. Société tunisienne de l'électricité et du gaz. (2016). Rapport annuel. Available on http://www.steg.com.tn/fr/institutionnel/publication/rapport_act2016/Rapport_FR2016.pdf

STEG. (Several years). Société tunisienne de l'électricité et du gaz. Rapport annuel.

Zekri, S., Echi, L., & Sghaier, M. (1996). Tunisia. In: A. Dinar, & A. Subramanian (Eds.), *Water pricing experiences, an international perspective.* (pp. 125–133). World Bank technical papers N° 386.

Mohamed Salah Bachta has graduated from INAT, Tunisia (1975), and CIHEAM IAM Montpellier, France (1976), with, respectively, BSc and MSc in Agricultural Economics. He earned his PhD in Agricultural Economics from the Université Catholique de Louvain, Belgium, in 1990. He worked at the National Center for Agricultural Studies (CNEA) for about 15 years, where he had conducted and led financial and economic analysis and feasibility studies for major irrigation and development projects in the country. He joined INAT (National Agronomic Institute of Tunisia), University of Carthage, in 1990, as Assistant Professor in the Department of Rural Economics. Since then, he had been promoted to Associate Professor and Professor in the same department. At INAT, he spent 25 years of active teaching, research, outreach, and consultancy work. He has worked extensively on agricultural and water economics issues in Tunisia. He led/coordinated a number of projects, namely, the PAR-PAA (Agricultural and Food Policy Analysis), MERGUSIE (Water Management of Merguellil River Watershed), WADEMED (Water Demand Management in the Mediterranean), etc. He supervised about ten PhD and several MSc students in areas of water management, agricultural, and environmental policy. He is Consultant to the FAO, World Bank, GIZ, and several other agencies.

Jamel Ben Nasr is Assistant Professor at the Carthage University, Department of Economics in National Agronomic Institute of Tunisia. His research interests include water management and governance, natural resource governance, institutions and agriculture performance, social capital, and rural development. He received his engineering degree in Agriculture and Rural Economic from the College of Agriculture, Mograne, Tunisia, in 2003, his MSc degree in Agricultural Policies and Agri-food Strategies from the National Agronomic Institute of Tunisia (INAT) in 2007, and his PhD degree in Agricultural Economics and Development from the National Agronomic Institute of Tunisia (INAT) in 2015. Between 2011 and 2012, he served as Senior Engineer at the Direction of Irrigated Areas at the CRDA Zaghouan (Tunisia).

Chapter 9
The Water Sector in MENA Region: The Way Forward

Slim Zekri

Abstract This chapter summarizes the most important water policies implemented in the seven countries considered in the book. It also assesses the progress made by each of the countries on the reforms judged fundamental to avoid water crises and social unrests and ensure water security for a sustainable development. The chapter concludes with a comparison of reforms undertaken worldwide and suggests some urgent measures that would speed up balancing supply and demand. Essentially demand management and cost recovery are judged urgent to undertake and feasible by learning from the successful energy price reforms undertaken in Iran and Oman for instance.

Keywords Water policies · Reallocation of water resources · Treated wastewater reuse · Cost recovery

9.1 Introduction

Traditionally, citizens of Middle East and North Africa (MENA) countries were wise users of water as witnessed by the historical water infrastructure and the way water was managed at homes and in public places in these arid countries (Salleh and Taher 2012). In the southern part of Tunisia, for instance, rainwater harvesting for drinking purposes was and still a must in most houses. Separation of grey water from bath rooms for garden irrigation was a common practice. The use of slightly saline water from wells for other uses than drinking was the rule as well as the use of dry latrines. People were paying the full cost of water and several charity institutions were in place to provide water for low-income families. With urbanization, centralized governments spread their water services to the communities and most of the aforementioned practices vanished, although an important part of the cost of

---

S. Zekri (✉)
CAMS, Department of Natural Resource Economics, Sultan Qaboos University,
Al-Khod, Sultanate of Oman
e-mail: slim@squ.edu.om

© Springer Nature Switzerland AG 2020
S. Zekri (ed.), *Water Policies in MENA Countries*, Global Issues in Water Policy 23,
https://doi.org/10.1007/978-3-030-29274-4_9

water services was and is still paid by public authorities, that is, sending distorted signals about water scarcity.

Generally speaking, MENA countries have not been able to accompany the social and economic changes with the required institutional and technical innovations in the water sector to cope with the water scarcity during the last 100 years. They contended themselves with the adoption of solutions designed to Western water-rich countries. It took long to properly realize the severe water scarcity in MENA countries and to start the journey for water policy reforms. Seminal works on MENA water issues date back to the 1990s. Gleick et al. (1994) addressed MENA water scarcity and potential wars on water as well as the potential solutions to avoid conflicts such as agreement on transboundary water, allocation of surface and groundwater rights, transfer of water from the agricultural sector to the urban sector, improving water use efficiency, desalinated water, water banks, and water markets, as well as the consideration of climate change effects. Brooks et al. (1997) summarized the way demand management can be undertaken as an essential and effective policy tool for Africa and the Middle East. They stressed on the following tools: institutions and laws, market-based measures, non-market-based measures, and direct intervention. The authors stressed that the conditions in the north are different to the point that MENA and African countries should rely on their own research to develop appropriate options. The Northern countries have utilized their capital assets and energy to overcome deplorably bad water management. Brooks et al. (1997) argued that solutions to water scarcity and conflicts should be looked from outside the water sector and wars over water are a myth. The authors introduced the concept of virtual water and international trade of agricultural products as a solution for water-short countries. Mubarak (1998) explicitly recommended that the agricultural and its role must shrink to avoid overuse and depletion of the water resources. The first World Bank (2007) report on water perfectly defined the problem and solutions to the region's water scarcity and brought the discussion to the political arena. The World Bank (2017) report is an update of the situation as well as the options available to decision makers. All recent and previous publications agree on the diagnostic and severity of the water situation in MENA and the need for reforms. In this chapter, a quick summary of the reforms that have been undertaken in the seven countries is presented with some recommendations for future reforms. Public decision makers and stakeholders of the water sector are called to perceive the overall picture of the water sector and to consider investments in the security and governance of water as investments in peace and security. There is an urgent need to consider a better allocation of the scarce water resources through demand side management such as water rights, regulations and quotas, water pricing, water trading, and subsidy reform (Cousin et al. 2019).

9.2 Water Policies in MENA

Flagship policies of the seven countries considered in this book are summarized in this section. Algeria created a Ministry of Water Resources since year 2000 with a mandate to elaborate and implement policies and strategies in the context of water resources and environmental protection. Algeria started since 2008 coordinating with its neighbors Libya and Tunisia to sustainably manage the common groundwater resource. Algeria also has agreements with Tunisia on the management of surface water in the Medjerda basin. In the Sahara, the Foggaras (equivalent of Aflaj in Oman) represent a good example of sustainable groundwater management. Some of these Foggaras are still alive and functional in several oases. However, in general, groundwater abstraction is still not monitored, though in some parts of the country groundwater rights are in use. The urban sector is given absolute priority of water allocation among sectors. The allocation to the agricultural sector is determined yearly after subtracting the demand for the urban sector. The supply and management of water utilities are achieved through a mixture of public and private entities. Although seawater desalination has been able to address the water shortages, it resulted in a considerable increase in energy use by the water sector, which reached 4983 GWh in 2011 and is set to more than treble to reach 16,090 GWh by 2030. The predicted increase is caused mainly by seawater desalination, water transfer projects, supply of water through pipes, and wastewater treatment facilities.

In Egypt, the water sector responsibilities are divided between two ministries: the Ministry of Water Resources and Irrigation, in charge of water resources planning and management, irrigation, and agriculture drainage, and the Ministry of Housing, Utilities and Urban Communities, in charge of domestic water supply and sanitation. The irrigation water delivery costs are indirectly recovered through agriculture land taxes. The costs of investments on irrigation improvement and drainage are recovered from the farmers in the form of installments. A large number of water user associations (WUAs) exist in Egypt and a new law aims to devolve more roles to the WUAs such as recovering the costs of irrigation, investment projects, and operation and maintenance of central pumps. Domestic water has the highest priority over other sectors in Egypt's water policies. In case of shortages, the domestic water demand takes priority in satisfying demand over other sectors. In 2018, urban water prices have been increased. Four block prices are used for domestic uses, while another four different prices are used for the different economic sectors. Wastewater treatment costs vary from 75% to 98% of the fresh water price. Private sector is being involved in wastewater treatment projects that treat and supply treated wastewater for reuse in the cities. Egypt depends almost totally on the surface water from the Nile River, whose basin is shared by 11 countries. Yet there is no formal agreement among all these countries on the water shares. However, a declaration of principles was signed in 2015 by Ethiopia, Sudan, and Egypt, agreeing on the rules of the first filling and the operation of the Grand Ethiopian Renaissance Dam.

In Iran, the water sector is under the responsibility of the Ministry of Energy (MoE) and the Ministry of Agriculture (MoJA). Both surface and groundwater uses are subject to water permit, which is issued for a specific use and nontransferable. Water is considered a national wealth, but groundwater is regarded as private property and thus some trade between farmers is allowed. The traditional Qanat (equivalent to Foggara and Falaj) system is also based on ownership of groundwater rights. Urban water prices are set at very low level that does not allow cost recovery for political reasons. Nonetheless, the government has put in place a national task force to address the problem of adaptation to water scarcity with demand management given the highest priority. The task force also extends to provinces where decisions on allocations, arrangements, and agreements between different sectors are reached on the basis of operational temporal and spatial demand. Mechanisms for settling water disputes are established at two different levels and are based on arbitration. Drought preparedness strategies for emergency situations are in place with established "Agricultural Product Insurance Fund," and extended loan reimbursement deadlines provision and aid measures such as the provision of agricultural inputs, small and medium funds, food subsidy, and forages in cases of drought/flood emergencies. The country has also finished comprehensive assessment work of global warming impact on the hydrology and water resources. One major policy to adapt to climate change consists of building desalination plants in the coastal cities in partnership with the private sector. Iran has treaties and agreements with most of its neighbors on the several Transboundary Rivers and shared wetlands. However, agreements on shared aquifers are lacking.

In Jordan, three organizations share the water responsibility. The Ministry of Water and Irrigation (MWI) is responsible for the policy making and sector planning. The Water Authority of Jordan (WAJ) is responsible for the infrastructure development and service delivery of both fresh and wastewater, and the Jordan Valley Authority (JVA) is in charge of the irrigation water in the Jordan Valley. Jordan has put in place an advanced groundwater policy. Groundwater is monitored and metered. Smart-card controlled abstraction and quotas are being introduced progressively in the wells. Farmers pumping more than 75,000 m^3/year pay a fee. Priority allocation of groundwater is given to high value uses, while supporting the sustainability of existing agricultural plantations. Expropriation of use rights arising is possible against fair compensation. Allocation of groundwater for irrigation is recommended from aquifers where water quality does not qualify for use in municipal and industrial purposes, and contingency plans shall be made for the purpose of allocating the water from privately operated wells for use in the municipal networks. Water user associations (WUAs) establishment started in 2001 with a diversity of legal status. However, most WUAs are legally under the cooperatives law and recognized by JVA. Treated wastewater is heavily used in agriculture and desalination of brackish water for irrigation is starting with 1.5% of total irrigation water being desalinated brackish water. Urban water services are intermittent and households depend on top roof reservoirs. Both fresh water and wastewater utility services are based on increasing three block tariffs. Users also pay an added value tax that is increasing according to the block tariff and it reaches $0.42 /$m^3$ for the upper block of above 100 m^3 per month. The total price of water services for the highest block tariff is quite high, for an intermittent service, with \$6.12 per m³. The lowest price is \$1.56 per m³ for demand below 15 m³/month. Assuming an average family of 5, this is the equivalent to 100 liters/cap/day at a low price that does not cover the cost of service. Cross subsidization is inherent in the block design. Jordan has bilateral water agreements with its neighbors to manage shared surface water resources in the Jordan basin.

In Oman, the responsibility for water legislation falls mainly in the hands of the Ministry of Regional Municipalities and Water Resources. However, since the most important part of urban water is from desalination plants, the Public Authority for Electricity and Water (recently renamed to Diam) and Oman Power and Water Procurement Company play a major role on policies related to this sector. The Ministry of Agriculture and Fisheries deals with policies related to the use of the water for irrigation. Finally, the water semipublic company Haya is the player when it comes to wastewater treatment and reuse. Most renewable water is in the form of groundwater. Water is declared a national wealth with the exception of the traditional Aflaj systems where water is still a private property that is exchanged in formal markets among farmers. Since year 2000, regulations have been issued to meter and monitor groundwater. Technical difficulties of metering have delayed the implementation of this crucial law and the trend today is to use smart metering technology for this purpose. Urban water prices are heavily subsidized despite the block tariff structure. Ninety percent of the supply depends on seawater desalination with a heavy reliance on fossil fuel for this purpose. Reuse of wastewater is almost confined to urban landscaping. Treated wastewater is owned by Haya Company and its high price prevents its use in the agricultural sector, given the open free access to groundwater. The impacts of climate change on the water sector are assessed with absence of action plans on adaptation and mitigation.

The Ministry of Environment, Water and Agriculture is responsible for policy and regulation in Saudi Arabia. The agricultural sector remains the principal consumer of water resources in the Kingdom. In the 1980s, the Kingdom started its very ambitious agricultural program to realize self-sufficiency to meet its food requirements. The goals were successfully achieved and the Kingdom witnessed self-sufficiency in many food commodities. Since 2008, the Kingdom decided to phase out the wheat production subsidy program, which was started in the 1980s, which allowed the Kingdom to become the sixth largest wheat exporter at the expense of its water resources. Meanwhile wheat was replaced by Alfalfa, which consumes three times more water than wheat. In 2015, the Saudi government decided to stop the cultivation of green feed within 3 years. Farmers would not be allowed to grow fodder after 2019 and are encouraged to produce less demanding water high value crops. The Ministry of Environment, Water and Agriculture is planning to impose metering on individual wells and to allocate water shares for users. Until 2015, consumers in Saudi Arabia paid only 1% of urban water service costs. In 2016, five block tariffs were introduced but prices are still on the low side and the water bill for consumption not exceeding 60 m³/month does not exceed \$24/month for water and sanitation. This will allow covering 30–35% of the costs and plans are in place to achieve full cost recovery.

The Ministry of Agriculture Hydraulic Resources and Fisheries is the main responsible for water in Tunisia with the exception for the treatment of wastewater, which falls under the responsibility of the Ministry of the Environment. The water law was promulgated in 1975, which established the ownership of all water resources as public resources, transforming old property rights into use rights. Resource allocation is done at central level through Master Plans and absolute priority is given to urban water. Since the 1990s the government started a vast reform program aiming to devolve the responsibility of water management to water user associations. Thousands of associations were created, but their dependence on the state is heavy due to political, technical, and financial interferences. Shallow groundwater is not monitored and illegal drillings are numerous with the consequences of overabstraction and salinization. Deep aquifers, whose exploitation is reserved to the public authorities, require much heavier investments and special equipment for drilling, which are not available easily. Treated wastewater reuse is very limited, given the low quality and availability of better surface/ground quality water. Disposal of secondary treated wastewater into the sea has severely degraded the quality of the beaches surrounding the capital city Tunis. A rigorous policy of annual increase of irrigation water prices has been implemented since 1991 in order to achieve cost recovery. WUAs have been authorized to set water prices taking into account the budget balance. Financial and technical incentives to save water are in place since 1995. The country depends on imports for staple food, while exporting water-intensive crops such as dates. Domestic water is metered and billed along with wastewater with cost recovery being a major policy. Users do even pay an added value tax on the water bill. Water allocation for the environment is well established even though during drought periods the primacy is for the urban sector. Tunisia has agreements in place with its neighbors Algeria and Libya on the use of the trans-boundary groundwater resources while for surface water only consultation mechanisms, particularly in case of flooding, are observed. Climate change impacts are well assessed, but action plans on mitigation and adaptation are not yet clearly planned.

9.3 Reforms Within the Water Sector

Table 9.1 summarizes the main reforms within the water sector in the seven selected countries. These reforms are related to the reallocation of water from rural to urban users and from agriculture to industry, freezing of the irrigated area, permitting and monitoring of groundwater and implementation of laws, reuse of the treated wastewater for agricultural purposes, tertiary treatment of the wastewater for food safety and environmental protection, cost recovery for urban and treated wastewater, and access of the poor to drinking water.

The reallocation of water from rural to urban users and from agriculture to industry has not been undertaken explicitly in any of the seven countries. Supply augmentation has been the rule so far. No clear-cut policies or water rights for the surface

9 The Water Sector in MENA Region: The Way Forward

Table 9.1 Degree of adoption of reforms within the water sector

	Algeria	Egypt	Iran	Jordan	Oman	Saudi Arabia	Tunisia
1. Any reallocation of water from rural to urban users and from agriculture to industry?	No	No	No	No	No	No	No
2. Are we still seeing an increase of the irrigated area?	Yes	Yes	Yes	No	No	Yes	Yes
3. Any allocation of quotas and monitoring of groundwater?	No	No	Yes	Yes	No	No	No
4. Tertiary treatment of wastewater	Yes	No	No	No	Yes	Yes	No
5. % of reuse of treated wastewater (TWW) in agriculture and landscaping	3%	7.2%	40%	91%	68%	50%	29.6%
6. Cost recovery for TWW	No	No	No	Yes	No	No	No
7. Cost recovery for urban water	No	Yes	No	No	No	No	Yes
8. Access of the poor/marginalized population to drinking water	Yes	Low	Yes	Yes	Yes	Yes	Yes

water exist on how much water should be allocated to the urban sector in the present or the future. However, except for Oman and Saudi Arabia, the urban sector has the highest priority when decisions of allocations are needed such as during drought periods. This implies that the urban sector is satisfied at the expense of the farming sector without any compensation for the later. One of the most important recommendations for the MENA region has been to cut down the irrigated area or at least not increase it in face of the increased competition for water. Five of the seven countries have been seeing an increase in the irrigated area during the last decade, except Jordan and Oman. It looks like that MENA countries are not being able to find alternative rural development options to the expansion of irrigated agriculture. Still the easiest option is to support farmers' access to irrigation water as a mean of income improvement and stabilization of the populations in the rural areas. Even though reduction of the irrigated area seems to be an internal water sector reform, in fact rural development based on industrial activities is a necessary reform that should come from outside the water sector to offer well-paid jobs to reduce the dependency on irrigated farming. Unfortunately, rural development is totally neglected in most of the MENA countries or development options are restricted to agricultural development (Nin-Pratt et al. 2018). Anticipating water depletion in rural areas is of extreme importance in several locations, mainly when unsustainable groundwater is used. MENA countries have to have clear plans on how much water would be sustainable in many villages depending on irrigation and prepare plans for relocation of the population or changing the economic activities to less dependent activities on water.

Groundwater reforms have been introduced in Jordan and Iran. The best example in MENA is Jordan with clearly established water rights and metering. Iran has water permits in place indicating how much water can be pumped from a given well in some areas, but no water rights. In the remaining MENA countries considered here, not only groundwater is still an open access resource, but wrong incentives such as subsidies to electricity and modern irrigation systems are causing further abstraction of the resource. Electrification of rural areas and availability of efficient submersible pumps have encouraged well owners to keep deepening their wells and abstracting more than the renewable rate of the resource. Decision makers are not interested in creating new laws and regulations to the management of the aquifers because this would entail more commitment and more work to be done administratively. Top decision makers are not willing to fund programs of property rights allocation and wells metering and monitoring due to the opposition of farmers. Furthermore, the existing technology of mechanical meters is useless and the smart online groundwater meters are still not affordable by nonoil countries (Zekri et al. 2017). Although some small communities have experiences of collective management, which is effective, there still remain the exceptions and cannot be generalized to aquifers tapped by hundreds of farmers. To be successful, groundwater policy reforms require that the online smart groundwater meters cost goes down, internet in the rural areas is generalized, and incentives to top decision makers to pass new laws and establish clear groundwater property rights are provided. The most challenging issue here is how to encourage public decision makers to undertake the reform. One option would be to fund such incentives by development aid institutions and link the payment to the achievement. Furthermore, some external support is recommended to establish pilot projects on groundwater smart metering and monitoring in MENA countries avoiding the overallocation of the resource and establishing high penalties for those who exceed their allocated quotas as main lessons from the Jordan experience.

Tertiary treatment of wastewater remains marginal in most countries except the oil-rich countries where most of the water treated goes to tertiary and even quaternary level. The environmental costs related to disposal of secondary treated wastewater are quite visible in the coastal areas where beaches have been polluted and are no longer swimmable. The question is whether MENA countries can afford investments on tertiary treatment. Most likely a direct payment for tertiary treatment is not affordable by the low- to middle-income families. The technology for tertiary treatment, in centralized plants, is still expensive for a wider adoption. Algeria has moved to tertiary treatment recently and has been able to bring back the capital city beaches to a clean level and usable for swimming and recreation. The consequent energy saving and time saving are substantial due to the opening of the capital city for swimming instead of travelling to nearby beaches for recreation. This positive externality would per se justify the partial funding of tertiary treatment by tax payers. An indirect payment, through beach front property taxes, can contribute to the funding of wastewater treatment plants and address partly the problem.

The reuse of TWW in agriculture and landscaping is progressing in MENA countries with Jordan on top of the list with 91% reuse rate. The high rate of reuse in Jordan is the result of important public investments to make the water available to farmers and the implementation of quotas for groundwater. Free access to groundwater does not encourage the reuse of treated wastewater. More public investments in MENA are needed to develop the necessary infrastructure to convey the treated wastewater up to the farms. Currently most treatment plants are connected to the sea to dispose of the water. One would ask who will pay for such costly pipelines for water transfer from the urban to the rural areas. It is obvious that farmers will not be able to fund water conveyance infrastructure given the low prices they are receiving for their agricultural products. They have not paid for water conveyance from dams in the past too. Private investors will not fund such projects because of the low profitability and long pay-back period. If food security is an objective, public investments can be tapped, since food security is a public good. However, public funds are scarce in most MENA countries and such an option is unlikely to happen. International funding institutions and donors should consider such funding under the climate change fund since treated wastewater can be used as an adaptation measure to drought.

Cost recovery for treated wastewater is not achieved in any of the seven MENA countries considered in this book. The same applies to domestic water cost recovery with the exception of Egypt and Tunisia. Tunisia has undertaken long ago a strategy to recover the full cost of urban water and was successful in achieving this goal. The prices remained low for the low-income families and a seven-block pricing method is used. In fact, the first three blocks' prices cover less than half the current cost of water and are meant to protect the low-income families. However, all customers pay a fixed cost plus a variable cost according to the block. Furthermore, all consumers pay an added value tax of 18% on the total water bill. Although the Tunisian case has been relatively successful in achieving cost recovery, the pricing method has not been updated or changed for several decades. Part of the cross subsidy (31%) is ending up in the pockets of wealthy families, given that the block prices do not take into account the number of family members. Tunisia is also facing expensive marginal cost of raw water paid in hard currency due to the recent adoption of desalination technology in several cities in the coastal areas.

In general, public water authorities in MENA should keep up with the reforms in the potable water sector and updating and upgrading the technology used in the distribution sector and provision of better water quality to citizens. Potable water price reform and pricing method should be a priority in the reform. Water consumption in several MENA countries is unreasonably high reaching +400 m^3/cap/day. Low administered water price is the main cause of such a behavior. Recent experience of the fuel price reform shows that people adjust properly to higher prices and no social uprising is mentioned, fundamentally when the low-income consumers are protected through direct subsidies. Oman introduced a fuel reform in 2016 by lifting the subsidy and establishing fuel price based on international market prices (Boughanmi and Khan 2019). The low-income families receive a direct subsidy, covering the price difference for 200 liters of fuel per month, every time the price goes beyond Omani Rials 0.180/liter. Low-income families apply for the subsidy online and receive an electronic card to access the subsidy in the gas stations. Substantial positive impacts have been observed and sale of petrol and diesel in Oman dropped by 6.2% and 7.2%, respectively, during 2016 and 2017 compared to a constant rise in fuel demand over the past years. People tended to reduce unnecessary trips and prices of fuel-inefficient cars dropped by more than 20% in both the new and used markets (*Muscat Daily*, January 23, 2017). Krane (2018) argued that MENA countries have launched reforms of energy subsidies driven by a number of converging trends such as fiscal stress from low world oil prices, international environmental pressure, and an untenable growth in domestic consumption of exportable commodities. Paradoxically, Krane mentions the escalating regional instability as a positive factor conducive to reforms as well. The successful experiences in energy prices reforms can be taken as good practices to encourage subsidy reforms in the potable and wastewater sectors. Ex ante studies on best pricing methods and subsidy targeting are highly needed in all MENA countries to inform decision makers on the potential gains and the range of maneuver to squeeze demand and delay costly supply projects. Price reforms should be accompanied by advice provision to users on the available technologies that allow water saving at homes such as lower shower heads, dual flush toilets as well as technologies for grey water treatment for garden irrigation. Separation of grey water from black water should become a building code for the new buildings with gardens. Rainwater harvesting should be encouraged too, whenever possible, to reduce run-off in the cities. Water demand management should become the major component of policy changes in MENA to curb water demand. This should be accompanied by improvements in water services, essentially better water quality wherever needed to increase the acceptance of stakeholders and users. An increase of water prices accompanied with a better quality of drinking water would result in net saving for most users who depend on bottled water.

Urban water is quite more expensive than agricultural water. Thus, from an economic point of view, saving urban water is as important as saving agricultural water. The increase in water desalination capacity in the MENA region has been dramatic during the last 5 years. Relying on water purification to satisfy future urban water needs is not sustainable given the energy and environmental impacts of desalination, unless the shift to renewable energies is taken seriously. Thus, further saving urban water demand can postpone heavy investments in desalination capacity.

Several cities around the world have introduced smart water meters. One advantage of the smart meters is the reduction of nonrevenue water (see South Africa's experience) besides the detection of postmeter leakage, which can result in 5–10% of a city's water demand reduction if fixed (Britton et al. 2013). Sønderlund et al. (2016) reviewed and analyzed 21 papers on the effect of smart metering and feedback given to consumers on reduction of water consumption. Overall, they found that feedback to consumers on their water consumption does change consumption practices. Water saving depends on the type of feedback information, frequency, and granularity. The feedback was more effective when it was combined with time variable water price and depends on baseline consumption. High consumers tended to reduce consumption more than low water users. Thus, a combination of smart water meters and a revision of water pricing and pricing methods should go together to address the future potable water demand in MENA.

Generally speaking, the seven countries have facilitated access of the poor/marginalized population to drinking water. A solidarity tax on urban water users is established in Morocco to support funding of projects aimed to improve access to potable water for rural households (Ait Kadi and Ziyad 2018). This is a good example that could be followed by other MENA countries, which would accelerate the water access of rural households. If such a tax is properly explained to urban water users, it will be easily accepted given the solidarity meaning of it. Rural water projects need not be in the form of pipes, given that houses are scattered. In many places, rainwater harvesting reservoirs are a good option. They can be equipped with small motor-pumps at house level and avoid the community management problems. Individual reservoirs of a capacity of 60 m^3 allow a water security of 50 liters/cap/day of very high quality.

9.4 Reforms Outside the Water Sector

9.4.1 Water–Food Nexus

In MENA countries, the food self-sufficiency goal has been replaced by food security goal. However, aggregated data still show that farmers are the major user of water. Despite MENA countries' reliance on heavy food imports, Jägerskog and Kim (2016) estimated that agricultural products are responsible for 92% of internal water footprint. Furthermore, most of the water used for food production in MENA is blue water (water from reservoirs and aquifers) as opposed to green water (rainwater). This is due to the total dependence on irrigation in many MENA countries and partial irrigation of some crops in other countries. Consequently, any water transfer from the agricultural sector to other sectors of the economy will have consequences on the rural communities, unless properly compensated through water markets and via allocation of treated wastewater to agricultural purposes. This is currently not the case since in most MENA countries the priority of fresh water allocation is given to the urban sector. This management practice has been the rule since during droughts the irrigation water valves are shut down without any compensation while cities' water is not affected. The option of transferring wastewater to rural areas to mitigate scarcity, however, might not be economically feasible in most countries, since the major volumes are generated in low coastal areas far from where needed for irrigation. The cost of infrastructure and pumping up hill is quite high. This is being aggravated with the further concentration of the population in the coastal areas.

Saudi Arabia and the United Arab Emirates, two of the most water-scarce countries, have been investing in farmland acquisition as a strategy to mitigate water scarcity, reduce risk from international food prices volatility, and reduce dependence on virtual water hegemons (Jägerskog and Kim 2016). Some of the acquisitions were in Egypt, an already water-stressed country, and where crop production depends totally on irrigation. Furthermore, most of the deals, which cover 4.3 million hectares, are in the Nile River Basin in countries such as Ethiopia, Sudan, and

Egypt with poor rural communities and low level of agricultural technology. This will exacerbate the competition for blue water in these countries. This has been stated by Antonelli et al. (2017), who stressed that current MENA imports depend markedly on water resources available in other countries that are not always water-secure countries. It would have been wiser to acquire land in agriculture rain-fed countries since trading green virtual water is more efficient than trading blue virtual water. While green water can be used only to support vegetation growth in agricultural and ecological systems, blue water can be used to meet other more productive uses including urban uses. Trading blue water that is embedded in low-value water-intensive irrigated crops is an inefficient use of globally scarce water resources and MENA countries can improve their water balance by importing such crops (Gilmont 2015). Last but not least, reduction of food waste in most MENA countries will reduce the food trade balance deficit and generate some financial savings to invest in the irrigation sector.

9.4.2 Water and the Media in MENA

The role of the media in informing the public and water users is fundamental for supporting policy changes. A website search proves the absence of studies in MENA countries on the role of mass media and social media to inform the users on the challenges and options available for a water secure future. Public/private televisions hardly ever talk about water problems and options. Social media is also absent and managers of water utilities seldom send messages except when interruption of service is planned. Media in MENA should be used intensively to communicate to/with the public of the required urgent reforms and the cost of status quo. Water has to become visible in the news, documentaries, and debates. More communication and accurate information should be circulated in social media if decision makers would like to find the necessary support from well-informed users. Quite often people assume that water users are not willing to accept higher bills but observed behavior shows that users do pay much higher costs to adjust individually than if collective solutions are envisaged.

9.5 Climate Change

The International Panel on Climate Change expects that water shortages will hit the MENA region the hardest. Not only rainfall will be lower but also higher evapotranspiration rates are expected leading to higher crop water requirements and lower yields. As an example, Tunisia's both irrigated and rain-fed cereals' yields would decline under different climate change scenarios between 5% and 11% till 2050 (Breisinger et al. 2013). Although the seven countries considered in this book have undertaken studies on the impact of climate change, mitigation and adaptation measures are still not well prepared. Tunisia, for instance, counts on further diffusion of water-saving technologies to save water. That is the only measure suggested in their strategy to cope with the change. This measure has shown its ineffectiveness in the past, as water saved is used by the same farmer to further intensification or extension of the irrigated area. No saving will happen in the absence of water quotas and administration of low irrigation water prices. MENA countries will have to find workable solutions to climate change such as water banking through aquifer storage and recovery during rainy years to face prolonged droughts, and introducing better engineered water-harvesting techniques in wider areas not confined to the traditional locations (Adham et al. 2016). Extending the irrigated area will not be possible due to water demand growth and expected lower supply as a consequence of climate change.

9.6 Experiences Outside MENA

Experiences outside the MENA region showed that there are several ways to undertake policy changes. Albiac (2017) provides an excellent review of experiences from Australia, the European Union, the United States, and Israel, which is summarized below.

Australia has chosen to introduce water markets and favor the transfers from rural to urban areas with the commitment of huge public funds in the form of subsidies to farmers to improve irrigation efficiency and to buy back water for environmental purposes.

Europe concentrated on water quality rather than water quantity since scarcity is not imminent, except in the South. Surface water pollution was brought down, via command and control instruments, by heavy investments upgrading the wastewater treatment plants, industrial innovation leading to phosphates free detergents and the nitrates pollution reduction via best farming practices, and identification of vulnerable pollution zones. Cost recovery of urban water directive was instituted in 2012. Most transfers of water from rural to urban areas are through either central decisions giving priority to urban water users or informal water markets among farmers. Finally, large subsidies are allocated to farmers to improve irrigation efficiency and reduce the nutrient and salt pollution as well as the reuse of treated wastewater.

In California, the United States, water markets were initiated since the 1970s. These markets play an important role in water transfers from rural to urban areas and represent 4% of the total volume of water used in a year. Water markets growth is hindered by the regulations dealing with third party effects and environmental concerns. Substantial public spending for nonpoint pollution abatement is used but not yet achieving the desired goals.

In Israel, central monitoring using a combination of command and control with economic instruments was the approach used. The government decides on the permits for both surface and groundwater allocations with a full network of metering and generalization of treated wastewater for irrigation. Water is priced for all uses and fresh water is reallocated to industrial and urban uses while bringing treated wastewater to farms. Initially urban users cross-subsidized irrigation water prices until 2016 when farmers started paying full financial cost of fresh water. Expansion of supply is based on desalination plants.

9.7 Conclusions

MENA countries have undertaken several efforts to improve water sector policies. The seven chapters show that the reforms vary considerably from a country to another. The major reform of proper water pricing and cost recovery is the most important policy in the water sector. Unfortunately, most countries considered in this book have not achieved this goal yet. Even the oil-rich countries, where the family's income is high enough, and are the ones who have the most severe scarcity, are still delaying water price reform. In most MENA countries, a bottle of water in a supermarket is more expensive than 1 m³ of piped urban water. Price reform is the most important policy since it will bring down demand, delay building costly desalination plants, and reduce the pressure on wastewater treatment plants. Reuse of treated wastewater in irrigation remains a challenge in MENA countries and will require substantial public investments in conveying that water to farms. Most MENA countries depend on imports of food and will see this dependency increasing in the future due to population growth and climate change. Agricultural productivity will have to grow significantly without using more water. Workable and more realistic adaptation solution to climate change needs to be invented. Water for the environment is still the last priority, given the absolute scarcity mainly during drought periods. World experiences of water policy reforms indicate clearly the necessity of public intervention and substantial spending to overcome the costs of water scarcity. A major part of the public funds currently directed to urban water service subsidies should be reallocated to undertake the necessary reforms in the water sector in the future. The current level of international aid to the water sector in MENA needs to be revised upward. It should also be better targeted and linked to achieve some desirable changes such as passing new laws and allocating water rights to different types of users, although command and control measures followed by some countries allowed reforming the water sector. However, establishing water rights and quotas is a better instrument in water-scarce countries and several MENA countries have had such mechanisms in the past or still have some working but not updated.

References

Adham, A., Wesseling, J. G., Riksen, M., Ouessar, M., & Ritsema, C. J. (2016). A water harvesting model for optimizing rainwater harvesting in the wadi Oum Zessar watershed, Tunisia. *Agricultural Water Management, 176*(2016), 191–202. https://doi.org/10.1016/j.agwat.2016.06.003.

Ait Kadi, M., & Ziyad, A. (2018). Integrated water resources Management in Morocco. Chapter 6. In *Global water security* (pp. 143–163). Singapore: Springer.

Albiac, J. (2017). *Review of the political economy of water reforms in Agriculture* (Working document 17/01). Department of Agricultural Economics. Agrifood Research and Technology Center. Zaragoza. Spain. 26 pages.

Antonelli, M., Laio, F., & Tamea, S. (2017). Water resources, food security and the role of virtual water trade in the MENA region. In *Environmental change and human security in Africa and the Middle East* (pp. 199–217). Cham: Springer International Publishing.

Boughanmi, H., & Khan, M. A. (2019). Welfare and distributional effects of the energy subsidy reform in the Gulf cooperation council countries: The case of Sultanate of Oman. *International Journal of Energy, 9*(1), 228–236.

Breisinger, C., Al-Riffai, P., Robertson, R., & Wiebelt, M. (2013). Economic impacts of climate change in Tunisia: A global and local perspective. In D. Verner (Ed.), *Tunisia in a changing climate: Assessment and actions for increased resilience and development* (pp. 59–77). Washington, DC: World Bank.

Britton, T. C., Stewart, R. A., & O'Halloran, K. R. (2013). Smart metering: Enabler for rapid and effective post meter leakage identification and water loss management. *Journal of Cleaner Production, 54*(2013), 166–176. https://doi.org/10.1016/j.jclepro.2013.05.018.

Brooks, D. B., Rached, E., & Saade, M. E. (1997). Management of water demand in Africa and the Middle East: Current practices and future needs. In *Introduction* (pp. 1–10). Ottawa: International Development Research Centre.

Cousin, E., Kawamura, A. G., & Rosegrant, M. W. (2019, March). *From scarcity to security: Managing water for a nutritious food future*. The Chicago Council on Global Affairs.

Gleick, P. H, Yolles, P., Hatami, H. (1994, April). Water, war & peace in the Middle East. *Environment, 36*(3), 6.

Gilmont, M. (2015). Water resource decoupling in the MENA through food trade as a mechanism for circumventing national water scarcity. *Food Security, 7*(6), 1113–1131.

Jägerskog, A., & Kim, K. (2016). Land acquisition: A means to mitigate water scarcity and reduce conflict? *Hydrological Sciences Journal, 61*(7), 1338–1345. https://doi.org/10.1080/02626667 .2015.1052452.

Krane, J. (2018). Political enablers of energy subsidy reform in middle eastern oil exporters. *Nature Energy, 3*, 547–552.

Mubarak, J. A. (1998). Mubarak Middle East and North Africa: Development policy in view of a narrow agricultural natural resource base. *World Development, 26*(5), 877–895. https://doi.org/10.1016/S0305-750X(98)00015-1.

Muscat Daily. (2017, January 23). Fuel sales drop as motorists shift focus to optimizing cost.

Nin-Pratt, A., El-Enbaby, H., Figueroa, J. L., ElDidi, H., & Breisinger, C. (2018). *Agriculture and economic transformation in the Middle East and North Africa: A review of the past with lessons for the future*. Washington, DC: International Food Policy Research Institute (IFPRI). orcid. org/0000-0001-9144-2127.

Salleh, S. A. A., & Taher, M. T. (2012). Rooftop rainwater harvesting in modern cities: A case study for Sana'a City, Yemen. *Journal of Science & Technology, 17*(2), 48–68.

Sønderlund, A. L., Smith, J. R., Hutton, C. J., Kapelan, Z., & Savic, D. (2016). Effectiveness of smart meter-based consumption feedback in curbing household water use: Knowns and unknowns. *Journal of Water Resources Planning and Management, 142*(12), 04016060.

World Bank. (2007). *Making the Most of scarcity accountability for better water management results in the Middle East and North Africa* (p. 270).

World Bank. (2017). *Beyond scarcity: Water security in the Middle East and North Africa* (p. 199). https://doi.org/10.1596/978-1-4648-1144-9.

Zekri, S., Madani, K., Bazargan-Lari, M., Kotagama, H., & Kalbus, E. (2017). Feasibility of adopting smart water meters in aquifer management: An integrated hydro-economic analysis. *Agricultural Water Management, 181*, 85–93. https://doi.org/10.1016/j.agwat.2016.11.022.

Dr. Slim Zekri is Professor and Head of the Department of Natural Resource Economics at Sultan Qaboos University (SQU) in Oman. He earned his PhD in Agricultural Economics and Quantitative Methods from the University of Cordoba, Spain. He is Associate Editor of the journal *Water Economics and Policy*. He has worked as a Consultant for a range of national and international agencies on natural resource economics, policy and governance, agriculture, and water economics in the Middle East and North Africa. He is Member of the Scientific Advisory Group of the FAO's Globally Interesting Agricultural Heritage Systems. His main research interests are water economics and environmental economics. In 2017, he was awarded the Research and Innovation Award in Water Science from the Sultan Qaboos Center for Culture and Science.

Index

A
Algeria, 2, 7, 9, 13, 14, 19–43, 178, 187, 190–192

C
Climate change, 4, 6, 20, 24–27, 33, 34, 39, 42, 43, 48, 75, 78, 79, 83, 88, 108, 142, 150, 178–180, 182, 186, 188–190, 193, 196, 198
Cost recovery, 11, 14, 26, 28, 50, 60, 89, 90, 96, 102, 124, 132, 148, 152, 169, 188–191, 193, 197, 198

D
Desalination, 3, 13, 20, 52, 76, 90, 114, 136, 174, 187

E
Egypt, 2, 3, 14, 47–61, 99, 187, 191, 193, 195

G
GERD, *see* Grand Ethiopian Renaissance Dam (GERD)
Governance, 24, 86, 114, 125, 147, 163, 181, 186
Grand Ethiopian Renaissance Dam (GERD), 56, 57, 187
Groundwater, 3, 20, 49, 67, 86, 114, 136, 166, 186

I
Institutions, 7, 26, 31, 41, 42, 48, 50, 59, 60, 75, 79, 81, 83, 86, 93, 97, 107–109, 114–116, 126, 143, 147, 154, 163–166, 169, 170, 181, 185, 186, 192, 193
Iran, 2, 3, 13, 63–83, 188, 191
Irrigation schemes, 26, 28, 51, 96, 97

J
Jordan, 2, 3, 7, 10, 11, 86–109, 188, 189, 191, 192

L
Low tariff, 15, 148

N
Nexus, 38–40, 54, 79, 105, 129, 153, 177, 195
Nile, 47–51, 53, 54, 56–61, 106, 187, 195
Nubian Aquifer, 55–58

P
Policy, 2, 8, 12, 13, 19–43, 47–61, 63–83, 86–109, 114–132, 146, 152–154, 162, 164, 166, 168–170, 186, 188–190, 192, 194, 196–198
Pricing, 11–13, 26–27, 31, 35, 58, 72, 77, 94, 96, 98, 115, 123–125, 144, 163, 169, 174, 186, 193, 194, 198

© Springer Nature Switzerland AG 2020
S. Zekri (ed.), *Water Policies in MENA Countries*, Global Issues in Water Policy 23,
https://doi.org/10.1007/978-3-030-29274-4

R

Reallocation of water resources, 58, 59
Research, 20, 27–29, 42, 43, 71, 72, 75, 79, 83, 93, 98, 101, 109, 125, 146, 149, 164, 180, 181, 186
Reuse, 10, 11, 20, 25, 26, 32, 34, 49, 53–55, 57, 58, 60, 61, 77, 94, 95, 103, 109, 118, 119, 122, 125, 142, 143, 187, 189, 190, 192, 197, 198
Ruthless extraction, 136

S

Scarcity, 2, 7, 10–12, 25, 29, 33–35, 40, 42, 67, 73, 78, 83, 86, 103, 106, 108, 120, 122, 131, 137, 140, 142, 147, 154, 162, 186, 188, 195, 197, 198
Sustainability, 2, 6, 23, 28, 47, 57, 93, 96, 98, 102, 108, 109, 118, 122, 125, 137, 147, 152, 154, 164, 176, 188
Sustainable agriculture, 121

T

Tariffs, 11, 15, 53, 60, 77, 101, 102, 104, 119, 123, 124, 136, 144, 146, 148, 149, 151, 169, 170, 174, 188, 189
Treated wastewater reuse, 10, 11, 96, 142

V

Virtual water, 29, 42, 51, 52, 99, 100, 116, 126, 173, 186, 195, 196

W

Wastewater, 3, 20, 49, 72, 86, 114, 168, 187
Water demand, 2, 5, 7, 12, 14, 26–29, 33, 42, 47, 48, 50, 51, 59, 61, 72, 98, 99, 109, 117, 121, 124, 132, 137, 140, 146, 147, 149, 151, 154, 169, 171, 172, 179, 180, 187, 194, 197
Water policies, 2, 5, 19–43, 47–61, 63–83, 86–109, 114–132, 153, 154, 163, 164, 186–190, 198
Water prices, 11, 12, 14, 28, 36, 77, 96, 98, 101, 123, 132, 149, 154, 169, 170, 182, 187–190, 193, 194, 197, 198
Water resources, 2, 19, 47, 64, 86, 114, 136, 162, 186
Water supply, 5, 7, 8, 12, 20, 26, 30, 33, 41, 42, 48, 51, 52, 73, 86, 91, 98, 99, 101, 102, 105, 107, 108, 116, 118, 132, 136, 138, 142, 143, 145, 152–154, 164, 166, 174, 176, 177, 180, 187
Water users, 12, 15, 50, 61, 70, 90, 97, 108, 115, 118, 145–147, 182, 187, 188, 190, 194, 196, 197